Respiration
in the
Invertebrates

MACMILLAN STUDIES IN COMPARATIVE ZOOLOGY

General Editors: J. B. Jennings and P. J. Mill, University of Leeds

Each book in this series will discuss an aspect of modern zoology in a broad comparative fashion. In an age of increasing specialisation the editors feel that by illustrating the relevance of zoological principles in a general context this approach has an important role to play. As well as using a wide range of representative examples, each book will also deal with its subject from a number of different viewpoints, drawing its evidence from morphology, physiology and biochemistry. In this way the student can build up a complete picture of a particular zoological feature or process and gain an idea of its significance in a wide range of animals.

Respiration in the Invertebrates

P. J. MILL, B.Sc., Ph.D
Lecturer in Zoology, The University of Leeds

MACMILLAN
ST. MARTIN'S PRESS

© P. J. Mill 1972

First published 1972 by
THE MACMILLAN PRESS LIMITED
London and Basingstoke
Associated companies in New York
Toronto Dublin Melbourne
Johannesburg and Madras

SBN 333 13443 5 (cased) 13711 6 (paper)

Library of Congress catalog card number 72-90019

Printed in Great Britain by
Butler & Tanner Ltd
Frome and London

To my wife, Gillian

Preface

The intention of this book is to provide the undergraduate with a comprehensive account of invertebrate respiration from both physiological and structural viewpoints; however, much information is included that will prove valuable additional reading for Advanced level students. The subject is treated in a comparative way, rather than by presenting the reader with a series of chapters working through each group of animals in turn.

All animals require a supply of energy, both for actual physical activity and for the synthesis of protoplasmic compounds from the raw materials provided by the breakdown of their food. In some animals, such as certain parasites, this energy is provided anaerobically, that is to say without the involvement of oxygen as a hydrogen acceptor. However, the vast majority obtain their energy mainly from aerobic processes requiring the utilisation of atmospheric oxygen.

We may consider aerobic respiration as being composed of several distinct phases:

   (a) The uptake of oxygen from the surrounding medium.
   (b) The transport of this oxygen to the tissues.
   (c) The production of energy (intracellular respiration).
   (d) The removal of waste products (e.g. carbon dioxide) from the tissues.
   (e) The elimination of these waste products from the animal.

The first chapter deals with general considerations, particularly the need for the development of specialised respiratory structures. The different structural devices involved in oxygen uptake and the elimination of carbon dioxide from the body are dealt with in the next three chapters. The transport of oxygen and carbon dioxide provides the basis for chapters five and six, while chapter seven is devoted to intracellular respiration. The final chapter is concerned with the nervous control of respiration. An appendix is provided with a synopsis of the classificatory system used; and suggestions for further reading are contained in the bibliography.

Acknowledgements

I would like to thank all those who have helped during the various stages in the preparation of this book. In particular I would like to thank my wife for her encouragement throughout this period and for correcting and commenting on the final text, and also Drs. E. Broadhead and J. B. Jennings who both read the first draft and made a number of valuable suggestions.

Several figures have been reproduced from other publications with the kind permission of the authors and publishers. They are as follows:

Figure 3·6, Dr. R. C. Newell and Logos Press; figure 4·6, Pergamon Press; figures 4·12, 4·13 and 4·15, Professor H. E. Hinton, the Marine Biological Association, the Royal Entomological Society, Academic Press and Cambridge University Press; figure 5·1, Dr. E. F. J. van Bruggen and Academic Press; figure 6·1, Drs. J. B. Jennings and R. Gibson and Pergamon Press; figures 8·1, 8·2, 8·4, 8·6, 8·7 and 8·11, Cambridge University Press; figure 8·12, Drs. T. Myers and E. Retzlaff and Pergamon Press; figure 8·13, Dr. R. D. Farley and Professors J. F. Case and K. D. Roeder and Pergamon Press; figure 8·14, Dr. R. D. Farley and Professor J. F. Case and Pergamon Press; figures 8·15 and 8·17b, Dr. P. L. Miller, Cambridge University Press and Academic Press; figures 8·16 and 8·20, Professor G. Hoyle and Pergamon Press; figures 8·17a and 8·18, Professor J. F. Case and Pergamon Press.

My thanks are also due to Drs. E. F. J. van Bruggen and R. E. Weber for sending me electron micrographs for figure 5·1.

Finally I wish to thank all those whose figures I have redrawn, the details of which will be found in the appropriate legends.

Contents

Contents

ONE
Introduction

In small animals simple diffusion of oxygen from the surrounding medium across the general body surface is sufficient for their aerobic respiratory requirements. This is possible for animals living in an aquatic, or at least a damp, environment; but for a terrestrial existence the permeable body surface necessary for such a mode of respiration would lead to rapid desiccation. However, as body size increases so general diffusion becomes progressively more inadequate for the respiratory needs of the animal. This is because its volume, in other words the number of respiring cells, increases more rapidly than its surface area, and it is this surface to volume ratio which provides one of the important limiting factors for this method of oxygen uptake.

For a comparatively small increase in linear dimensions the surface/volume ratio shows a marked decline, and this is particularly so if the increase is equally distributed in all dimensions. For example, a cube of 1 cm side has a surface/volume ratio of 6 : 1. If the volume is increased to 27 cm³ by an equal increase in all dimensions to produce a cube of side 3 cm, the surface/volume ratio falls to 2 : 1 (figure 1·1). However, if the increase is in only one dimension, producing a tube 27 cm long with a cross-section of 1 cm square, then this ratio only falls to 4.1 : 1. To follow this line of argument further, progressive increase in all dimensions will cause the surface/volume ratio to approach zero, whereas increase in only one dimension will impose a limiting value dependent on the values of the two fixed dimensions. In the above example this limiting value is 4.0; for a cross-section of 2 cm square it is 2.0. (The same ratios incidentally hold good for spheres and cylinders of corresponding dimensions.) Thus it would appear advantageous for an animal to increase in size by an increase in only one dimension. Alternatively, an increase in two dimensions to produce

I

a large, flattened disc-shape will also produce a limiting value for the surface/volume ratio, although this will be lower (2.0 if the fixed dimension is 1 cm). Furthermore, increase in only one or two dimensions would ensure that any given cell was never too far removed from the surface of the animal in terms of the length of the diffusion path which the oxygen would have to

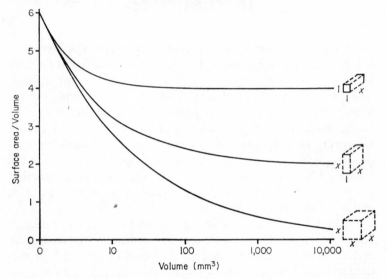

FIGURE 1·1 The relationship between the surface: volume (S/V) ratio and volume (V) when increase in size is in one, two or three dimensions

take to reach it, whereas if the increase were in all dimensions some cells would be a considerable distance from the surface.

The relationships between surface area, volume and metabolism can be derived from the theoretical consideration of a uniform body. Thus for a sphere or cube the surface area S can be expressed as a power function of the volume V (or weight W in a uniform body, where weight is directly proportional to volume). Thus, for a cube of side x units

$$S = 6x^2 \qquad (1)$$
$$V = x^3 \qquad (2)$$

From (2)

$$x = \sqrt[3]{V} = V^{1/3}$$

Substitute for x in (1)

$$S = 6(V^{1/3})^2$$
$$= 6V^{2/3}$$
$$= 6V^{0.67}$$

Or, in general terms

$$S = kV^b \tag{3}$$

In terms of weight rather than volume, k will have a different value, say k_a, which is partially dependent on the density of the body. Thus

$$S = k_a W^b \tag{4}$$

The value of k will vary according to the relative dimensions of the body.

Metabolism is generally measured in terms of the animal's oxygen consumption, although carbon dioxide or heat production would serve equally well. It may be referred to the oxygen consumption of the animal as a whole in terms of oxygen consumption/unit time ('Total Metabolism') or the animal's weight may be taken into account, i.e. oxygen consumption/unit weight/unit time ('Metabolic Rate'). The level of metabolism depends on a number of factors and varies considerably with the level of activity of the animal. Thus measurements are often stated in terms of a 'Basal' level, at which the animal is exhibiting no movement at all; a 'Standard' level, when there is a minimum level of movement; or an 'Activity' level, when the animal is exhibiting some predetermined level of activity such as walking, flying or swimming at some specific speed.

The relationship between body size and oxygen uptake is somewhat complex and varied. In general a large individual has a higher total metabolism than a small individual of the same species, but its metabolic rate is lower. The total metabolism M can, like surface area, be expressed as a power function of body weight W, since experimental data indicates a linear relationship between $\log M$ and $\log W$ (figure 1·2). Thus

$$\log M = \log k_c + b \log W$$

and hence

$$M = k_c W^b \tag{5}$$

The constant b in this case defines the rate at which oxygen consumption varies with size, and differs not only from one

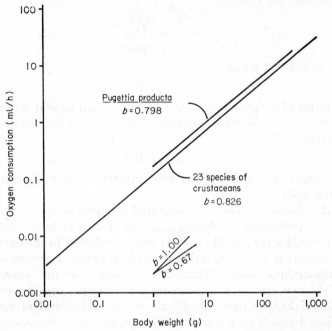

FIGURE 1·2 The relationship between total metabolism (measured as oxygen consumption/unit time) and body weight in different individuals of the kelp crab (*Pugettia producta*) (intraspecific) and in 23 different species of crustacean (interspecific). Double log plot. [After Weymouth, Crismon, Hall, Belding and Field (1944) *Physiol. Zool.*, **17**]

species to another but also with physiological circumstances and with the age of the individual. In some animals b approximates to 0.67 and so metabolism is directly proportional to surface area, since from (4) and (5) we have

$$S = k_a W^{0.67} \quad \text{and} \quad M = k_c W^{0.67}$$

Thus
$$\frac{M}{S} = \frac{k_c}{k_a}$$

$$M = S \cdot \frac{k_c}{k_a} \tag{6}$$

In other animals b equals 1.0 and so metabolism is directly proportional to weight, since from (5)

$$M = k_c W^1$$
$$= k_c W$$

In many other animals intermediate values occur.

One way in which this mathematical model can be used is to determine if any correlation exists between the value of b and the type of respiratory mechanism involved. Indeed it has been suggested that in those animals which possess gills the oxygen consumption is proportional to the surface area of the individual ($b = 0.67$); in animals with tracheae or lungs it is proportional to weight ($b = 1.0$); and in those which utilise two different methods such as tracheae and gills intermediate values obtain. A number of animals do in fact fit in with this arbitrary rule, but it is only fair to point out that there would seem to be more exceptions than animals for which it holds good. Another important consideration is that the more active the species the greater are the energy requirements and hence the greater the need to improve the efficiency of oxygen uptake.

Data are often presented in terms of metabolic rate R (figure 1·3a) and equation (5) may be rewritten for metabolic rate rather than total metabolism by dividing each side by W. Thus

$$\frac{M}{W} = k_c \frac{W^b}{W}$$

That is

$$R = k_c W^{(b-1)} \quad \left(\text{since } \frac{1}{W} = W^{-1}\right)$$

and hence

$$\log R = \log k_c + (b - 1) \log W$$

Since b is normally less than 1 these equations are generally written

$$R = k_c W^{b'} \tag{8}$$

and

$$\log R = \log k_c + b' \log W \tag{9}$$

where $b' = b - 1$. This also means that the linear relationship between $\log R$ and $\log W$ will have a negative slope (figure 1.3b). If metabolism is directly proportional to surface area b

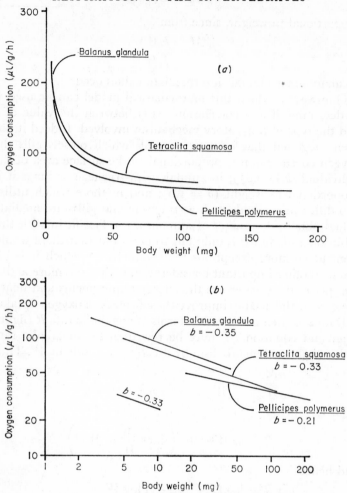

FIGURE 1·3 The relationship between metabolic rate (measured as oxygen consumption/unit weight/unit time) and body weight in three genera of cirripedes (barnacles). (a) linear plot; (b) double log plot. [After Barnes and Barnes (1959) *Veröff. Inst. f. Meeresforsch. Bremaerhven*, **6**]

will have a value of 0.67, and so b' will equal $0.67 - 1.00$, that is $- 0.33$.

We may now consider the environment of the animal in rather more detail. The transition to a truly terrestrial existence normally brings with it the problem of conserving water, and thus renders a generally permeable skin a hazard to survival rather than an asset. Restriction of oxygen uptake (and carbon dioxide loss) to certain specific and limited areas of the body surface permits the evolution of an impervious body wall. This has an advantage even in aquatic animals in that it allows a greater degree of ionic isolation from the environment than is otherwise possible. Alternatively one may argue that such restricted respiratory areas arose as a result of the development of an impervious body wall. No doubt both were of importance during the course of evolution, but nevertheless a relatively impervious body wall is an essential prerequisite for a truly terrestrial existence because of the attendant problem of water loss in such an environment. The most successful invertebrates in this respect, the insects, combat the problem with a very sophisticated and impermeable cuticle deposited by, and on the outside of, the epidermal cells.

In spite of these problems of dessication and respiration, there is one enormous respiratory advantage of a terrestrial existence, namely the use of air as a respiratory medium. Air contains approximately 25 times as much oxygen per unit volume as does water. (Dry air contains 20.95 per cent oxygen and exerts a pressure of 159 mm mercury at sea level.) Also the diffusion rate of oxygen is some three million times greater in air than in water. On the other hand carbon dioxide has a higher diffusion rate and is more soluble in water than is oxygen. These factors, coupled with the much greater density of water, mean that in those animals which have specialised respiratory structures a great deal more work must be done by an aquatic animal to pump the medium over its respiratory surfaces, and a high level of utilisation of the available oxygen is extremely important.

Another important variable is the oxygen content of water, which decreases with both increase in salinity and increase in temperature. This is shown in the accompanying table.

| | Oxygen content (ml/1 of water) | |
Temperature	Fresh water	Sea water
5°C	9.22	6.89
20°C	6.51	5.05

It is of interest to note at this juncture that an animal living in the intertidal zone has a particular respiratory problem, since the osmotic stresses to which it is subject require the expenditure of a considerable amount of energy to maintain some constancy of the body fluids. This all tends to support the generally accepted theory that the evolution of terrestrial life has generally occurred from sea water via fresh water, since the initial step would bestow a respiratory advantage on its perpetrators; whereas the osmotic stresses of an intertidal existence create a disadvantageous respiratory situation.

Among animals which live in an environment, such as the aquatic one, where environmental oxygen tension can be variable some, known as 'conformers', have a metabolic rate which steadily declines with increase in oxygen tension below air saturation; others, known as 'regulators', are able to maintain their metabolic rate under such conditions until a critical level of oxygen tension (p_c) is reached, below which they also become conformers. At the other end of the scale the metabolic rate of regulators remains steady with increase in oxygen tension above air saturation while that of conformers rises until some level is reached at which they in turn start to regulate. The tension at which this occurs can also be referred to as the critical level (p_c) but is often not very precise. Thus the position of the critical level above or below the normal environmental oxygen tension at air saturation determines whether an animal is classified as a conformer or a regulator. This is illustrated in figure 1·4 for different mayfly larvae in water at a temperature of 10°C, at which temperature the oxygen content of the water in equilibrium with air is 7.9 ml/l of water. Small larvae of *Baetis* for instance are conformers, with $p_c = 12.5$, while *Leptophlebia* $(p_c = 2.5)$ and *Cloëon* $(p_c = 1.5)$ are both regulators. Intermediates do occur, such as the larvae of *Ephemera*, in which there is no obvious critical level. Also there is an indication that some animals may change from conformers to regulators as they get older. Thus large larvae of *Baetis scambus* are regulators with their $p_c = 5$. *Leptophlebia* and *Cloëon* both live in fairly static water where an ability to regulate metabolism is very desirable, since fluctuations in oxygen tension are to be expected; *Baetis*, however, lives in streams and does not normally encounter low oxygen tensions. While a similar en-

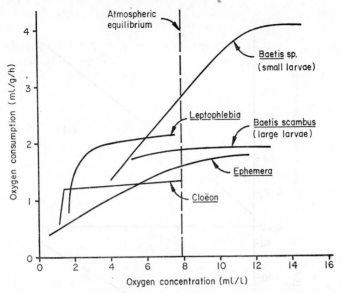

FIGURE 1·4 The relationship at 10°C between metabolic rate (measured as oxygen consumption/unit weight/unit time) and environmental oxygen concentration in ephemeropteran larvae.
[After Fox, Wingfield and Simmonds (1937) *J. exp. Biol.*, **14**]

vironmental pattern holds for many other animals, the reverse is often encountered. Also parasites which can respire aerobically tend to be conformers even though they often encounter low oxygen tensions.

One very important environmental factor is temperature. A rise in temperature causes the metabolic rate to increase up to some critical level, and the Q_{10} for this (i.e. the factor by which the metabolic rate increases for a 10°C rise in temperature) is often as high as 2 or 3. Furthermore, this may be associated with an increase in other processes, such as the rate of ventilation, and in an increase in the percentage of oxygen extracted from the environment (figure 1·5). In a number of animals the effect on metabolism of seasonal variations in temperature is counteracted to some extent by their ability to acclimatise.

As soon as the size of the animal or the nature of its environment precludes diffusion as the sole means of oxygen uptake some other way of coping with the cellular requirements must

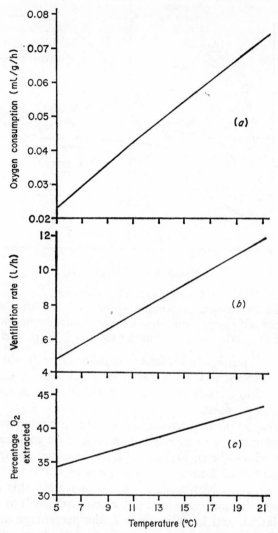

FIGURE 1·5 Effect of temperature on (a) oxygen consumption (uptake by the gills), (b) rate of gill ventilation and (c) percentage of oxygen extracted by a 345 g lobster in sea water with a uniform oxygen concentration of 5.3 ml/l. [After Thomas (1954) *J. exp. Biol.*, **31**]

be developed. The obvious answer, or at least the one which has been adopted, is to limit the site of oxygen uptake to a specific region or regions of the body and to increase the surface area of these regions without materially affecting the volume of the animal. Essentially three types of 'respiratory organ' have evolved—Gills, Lungs and Tracheae. All three meet the requirements of increasing the surface area of the body without any marked effect on its volume, and a detailed consideration of each of these structures is given in the following chapters. These specialised structures provide discrete areas of oxygen uptake. The first two, gills (which are normally only efficient in an aquatic environment) and lungs (which have evolved largely as an adaptation to a terrestrial existence), require the development of an efficient circulatory system to remove oxygen from its site of uptake and transport it to the cells in order to be effective. This transport system may be aided by the development of respiratory pigments, which enable more oxygen to be carried per unit volume of blood. A tracheal system (also a terrestrial adaptation) completely differs from gills and lungs in principle, in that atmospheric oxygen is normally taken into the close proximity of the cells by invaginations of the body wall. Thus the diffusion path to the cells may be extremely short and an efficient circulatory system is not really needed as far as the distribution of oxygen to the tissues is concerned.

All of these systems, to reach a high level of efficiency, necessitate the development of a pumping (ventilatory) mechanism to move the air or water over the respiratory surfaces in order to keep an oxygen-rich medium in contact with them. The amount of energy expended in removing oxygen from the surrounding medium and transporting it to its cellular destination must be minimal, compared to that ultimately made available by intracellular respiration as a result of this transfer. Regulators can maintain their metabolic rate in conditions of decreasing oxygen tension by increasing their ventilation or increasing their uptake of oxygen at the respiratory surface (withdrawal) or both; under similar conditions conformers show a decrease in ventilation or withdrawal of oxygen or both.

The development of respiratory organs does not necessarily mean that diffusion of oxygen via the general body surface is completely replaced and indeed many animals, principally aquatic ones, which possess such refinements still take up part of their oxygen in this manner. Furthermore the development of a circulatory system and the presence of respiratory pigments will help to improve the efficiency of general surface respiration in the absence of any discrete respiratory structures.

Concurrent with the problems of oxygen uptake are those of the disposal of the waste products of the respiratory process, principally carbon dioxide. The above arguments regarding size, etc., also apply in this respect. In a system utilising gills or lungs, most of the carbon dioxide is removed via these organs and again the importance of efficient circulatory and ventilatory systems is evident; but in an animal possessing tracheae a large proportion of the carbon dioxide may be removed over the non-sclerotised portions of the body wall rather than via the tracheal system, and some form of circulatory system would presumably be helpful in this respect also.

Some animals, notably endoparasites such as certain trematodes and nematodes, only respire anaerobically, and in these groups mechanisms have been evolved to deal with some of the rather toxic end-products of the process.

Further ways of improving the uptake of oxygen and disposal of carbon dioxide include the development of a feedback system whereby ventilation and/or circulation are increased under conditions of respiratory stress, and the development of a countercurrent system whereby the respiratory medium and the circulatory fluid are moved in opposite directions where they come into juxtaposition at the respiratory surfaces. This latter development is most advantageous when the velocities of the respiratory medium and the circulatory fluid are equal, and becomes less so as the disparity between the velocities increases (figure 1·6).

The only group of animals below the true coelomates to show any specialised region which may be concerned with oxygen uptake are the nemertines. Some of these animals actively pump water in and out of their foregut, which is highly vascularised. The nemertines are also the first group to demonstrate a circulatory system which is at least in part closed, and the circulatory fluid (blood) contains various coloured pigments in different species, a few possessing haemoglobin. The other acoelomate phylum, Platyhelminthes, also contains a few species (some rhabdocoeles) which possess haemoglobin in the mesenchyme, but it is uncertain whether it has any respiratory function here.

Among the pseudocoelomates no phylum contains species possessing any specialised structure for oxygen uptake, but

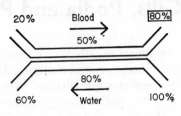

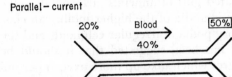

FIGURE 1·6 Oxygen uptake at the respiratory surface with (*a*) parallel (*upper figure*) and counter flow (*lower figure*) of the blood and the respiratory medium. [Based on Hughes (1963) *Comparative Physiology of Vertebrate Respiration*. Heinemann: London]

some of the nematodes at least contain a small amount of haemoglobin in their perivisceral fluid. It is only when we reach the level of organisation of the coelomates on the one hand and the deuterostomes on the other that we find phyla in which complex modifications of the body surface associated with oxygen uptake have occurred, and it is with these animals that we will be primarily concerned in the ensuing chapters.

TWO
Gills, Podia and Papulae

Gills are to be found in only four invertebrate phyla: annelids, arthropods, echinoderms and molluscs. In some hemichordates and in protochordates gill slits are present, but no gills. The term 'gill' can be applied to any outgrowth of the body wall that is primarily concerned with increasing oxygen uptake from the surroundings, and includes in this context not only the serially repeated gills of annelids, arthropods and certain molluscs and the ctenidia of the higher molluscan classes, but also the respiratory podia of irregular echinoids and the papulae of asteroids. It has been suggested that it should be limited to those structures which take up more oxygen per unit area than the general body surface does. It could be asked at this point whether increase in surface area or increase in oxygen uptake per unit area came first, but since this may vary in different animals it would seem more logical to consider the primary function of an organ, rather than the course by which this function has been achieved. It does not matter from the point of view of an animal's efficiency whether a small gill is developed which effectively takes up more oxygen per unit area or whether a larger gill is developed whose oxygen consumption per unit area is the same as that of the general body surface. In both cases the net effect is similar. The way in which a gill has evolved must have been dependent on the selective advantages offered by the various alternatives. Presumably all have evolved in response to respiratory stress. This may result in the formation of new structures, exclusively or primarily for respiration from the start, or an existing structure may take on the function of a gill. The respiratory podia of irregular echinoids are an example of the latter. Podia are thought to have been evolved primarily for movement, and are still used for this purpose in most instances, although admittedly they may always have served in a respiratory capacity. Similarly, it is generally accepted that the hemichordate and protochordate gill slits, which are an important site of respiratory exchange, developed originally to provide exhalant apertures for the food current entering the mouth; but again they have probably always had some respiratory importance.

Gills of many morphologically different types occur, ranging in form from simple plate-like extensions of the limbs to the complex ctenidia of higher molluscs. Indeed there are so many different forms of gill that simple gills such as gill plates and filaments and those with simple branching will be dealt with first, and will be followed by an account of the more complex, multiple or branched gills. Podia and papulae are included under the first heading.

Finally there are a number of animals which possess tentacles which, although not having a primary respiratory function, nevertheless are important as sites of oxygen uptake simply because of the large surface area they provide compared with the rest of the body. Included here are the lophophorate coelomate phyla (Phoronida, Ectoprocta and Brachiopoda), the ectoprocts and the pogonophores, and also some annelids.

The gills of most animals contain haemolymph, either because they are in direct continuity with the coelomic cavity or because they contain branches of the circulatory system. There is one group, however, which is an exception to this general rule, namely the insects, the gills of which contain air-filled tubes called tracheae.

SIMPLE GILLS

Simple gills are often arranged segmentally, one pair to each segment, over a few or many segments. This situation is probably found in most animals but there are a great many other gill arrangements. For instance there may be only a single pair, as in the ostracods (Crustacea), or they may be confined to the posterior end of the animal, as in the larvae of damselflies (Insecta), and so on.

2.1. Annelida Gills are to be found in representatives from all three classes of this phylum—Polychaeta, Oligochaeta and Hirudinea—although it is only in the first of these that they can be said to be of common occurrence. Polychaetes are divided into three groups, only two of which concern us here. These are the errant polychaetes, which includes free-living species, and the sedentary polychaetes, which includes tube-dwellers and those which live in permanent burrows. However, the distinction is not quite so sharp as appears at first sight since some errant polychaetes do in fact burrow and some even construct tubes.

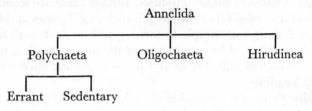

In general the segments of errant polychaetes are all similar and the lateral outgrowths of the segments (parapodia) are well developed. Gills, when present, are borne on most segments except at the extreme anterior and posterior ends of the body. Typically the parapodia each consist of a dorsal (notopodial) and a ventral (neuropodial) lobe, either or both of which may be further subdivided (figure 2·1). Also associated

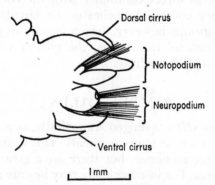

FIGURE 2·1 Anterior view of a parapodium of *Perinereis cultrifera*

with each parapodium is a dorsal and a ventral cirrus. Simple gills are derived either from one or more of the parapodial lobes, from the dorsal cirrus, or from an outgrowth at the base of the latter. Thus in *Nereis* the dorsal lobe of the notopodium

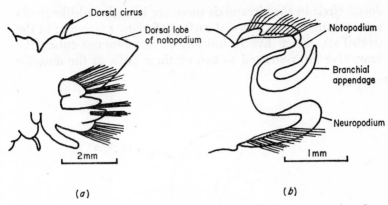

Figure 2·2 Gills developed from the parapodial lobes: anterior view of the parapodia of the errant polychaetes (a) *Nereis diversicolor* and (b) *Nephthys cirrosa*

acts as a gill (figure 2·2a), and in some other nereids and in *Nephthys* the ventral lobe of the neuropodium also serves the same purpose; furthermore the notopodium in *Nephthys* is extended to form a branchial appendage (figure 2·2b). In the phyllodocids it is the dorsal cirrus that has become expanded to form a large flattened branchial lobe, as in *Phyllodoce*, for example (figure 2·3a). The Eunicidae possess a variety of branchial structures arising from the base of the dorsal cirrus: in *Eunice* itself the gills may be simple, but are usually pectinate (figure 2·4a), while in *Marphysa* they are simple or branched (figure 2·4b). *Nothria* has a cirriform branchial projection from the base of the dorsal cirrus, but in *Diopatra* this structure is spirally branched. *Diopatra* is one of the tubiculous errant poly-chaetes and the gills are confined to the anterior few segments behind the head.

In the sedentary polychaetes the body often shows some degree of regional differentiation, the parapodia are poorly developed, and the gills are restricted to comparatively few segments. As in errant polychaetes the dorsal cirrus is often adapted to form a branchial lobe. This may take the form of a long inverted cone, as in *Sabellaria* and *Nerine* (figure 2·3b), or it may be a more complex structure with irregular branched filaments, as in *Arenicola* (figure 2·3c). Various other branchial structures occur that are not associated with the parapodia or

dorsal cirri: in the cirratulids there are long thread-like structures arising from the dorsal surface of many segments; in the
terebellids, which live in mucous or membranous tubes, the
branchiae are restricted to two or three pairs at the anterior

ERRANTIA

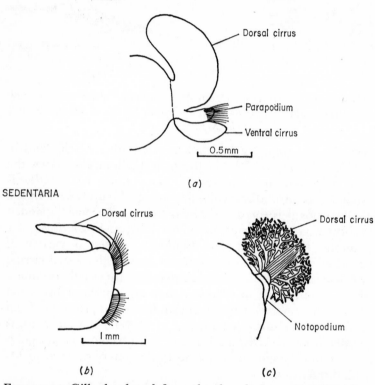

SEDENTARIA

FIGURE 2·3 Gills developed from the dorsal cirrus: anterior view
of the parapodia of the polychaetes (a) *Phyllodoce,* (b) *Nerine foliosa*
and (c) *Arenicola marina.* [c after Wells (1944) *Proc. Zool. Soc. Lond.*
114]

end. Again they are dorsal structures and are normally arbor-
escent, as in *Lanice, Loimia* and *Amphitrite,* but may be filiform
or pectinate. Finally, in the tube-dwelling fanworms (Sabellidae
and Serpulidae) there is a terminal crown of branchiae.

In the Oligochaeta, which includes earthworms, comparatively few species possess gills but they are to be found in some aquatic members of the class, several species of which possess tubular, segmental gills. Thus in *Branchiodrilus* there is a simple anterodorsal pair on all the body segments except those at the two extremities of the body, with the more anterior ones longer than the others. *Branchiura sowerbyi* has a dorsal and a ventral gill on each of the more posterior segments, and in *Hesperodrilus branchiatus* there are 13 pairs of lateral gills, again on the more posterior segments. Two species of *Alma* also have gills on the more posterior segments. These comprise from one to several small cylindrical projections in a transverse row on the dorsal

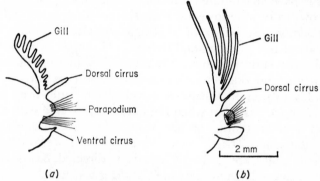

(a)   (b)

FIGURE 2·4 Gills developed from the base of the dorsal cirrus: anterior view of the parapodia of the errant polychaetes (a) *Eunice harassii* and (b) *Marphysa sanguinea*. [a after Eales (1952) *The Littoral Fauna of Great Britain*. Cambridge Univ. Press]

surface on each side of the mid-line. They may be simple or branched. In a third species of *Alma* they are restricted to the margin of the anus and do not branch. In *Dero* and *Aulophorus* there is a depression called the branchial fossa at the hind end of the body, and two to four pairs of contractile tubular or plate-like gills arise from the inner wall of this depression or from its margin.

Similarly in the Hirudinea (leeches) gills are uncommon and only two or three species of piscicolids, the so-called fish leeches belonging to the order Rhynchobdellida, possess them. *Branchellion*, which is parasitic on various teleosts and elasmobranchs, has 31–33 pairs of lateral foliaceous branchiae (three pairs per somite). *Ozobranchus*, a parasite of chelonians and crocodiles, has a single pair of finger-shaped branching gills on each abdominal somite—*O. jantzeanus*, for example, having 11 pairs.

Respiratory currents. Respiration is aided in many polychaetes by ventilation of the gills by a respiratory current. There are basically two ways in which such currents are produced: by the beating of cilia and by muscular movements. Many species have cilia on the gills and on the dorsal surface of the body and these beat to produce a backwardly directed flow of water. *Harmothoë* produces its respiratory current in this way, as does the tube-swelling *Nerine*.

It is particularly important for animals living in tubes and burrows to be able to renew their surrounding water. Thus in *Arenicola marina* bursts of rhythmic activity alternate with periods of rest. This animal lives in a U-shaped tube and the respiratory current is anteriorly directed. When the burrow is exposed at low tide *Arenicola* ceases to irrigate its burrow, presumably to prevent warm, oxygen-depleted water from being drawn in (p. 124), but when the burrow is covered with water again hyperventilation occurs. Many other tube-dwellers, such as *Eupolymnia*, *Thelepus* and *Neoamphitrite*, have respiratory currents which are primarily anteriorly directed. *Sabella*, however, is somewhat unusual; in *S. spallanzanii*, which lives in a U-shaped tube, the current may be in either direction, but in *S. pavonina*, whose tube terminates well below the surface, it is normally posteriorly directed. The errant burrowing worm *Glycera* also uses muscular movements to ventilate its gills.

Respiratory currents produced by muscular movements, however, are not the sole prerogative of tubicolous and burrowing species, and *Nereis* is an example of a free-living species that uses undulatory body movements rather than cilia for this purpose. With regard to oligochaetes and hirudineans, relevant information appears to be lacking, except in tubificids (e.g. *Tubifex*), which wave the posterior end of their body in the water. *Aulophorus* and *Dero* at least have cilia on their gills, but whether or not they are used to produce a respiratory current is not readily apparent. In *Branchiura* the gills move as well as the whole posterior end of the animal. The leech *Ozobranchus* has pulsatile gills, but the most highly developed respiratory system is found in the sea mouse *Aphrodite* (Polychaeta). Each notopodium bears numerous chaetal threads which are inter-woven over the dorsal surface of the animal. Water is drawn into the space between the dorsal body wall and the matted chaetae, and in this space the elytra (plate-like modifications of the dorsal cirri) are tilted in a regular sequence to produce a backwardly directed flow of water. It has been suggested that the elytra themselves act as gills.

2.2. Crustacea Gills are to be found in four of the eight crustacean subclasses—Branchiopoda, Ostracoda, Cirripedia and Malacostraca—and in all cases they are associated with

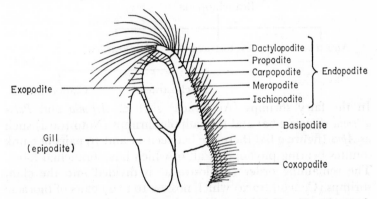

FIGURE 2·5 First thoracic appendage of *Nebalia* (Leptostraca). [After Calman (1909), in *A Treatise on Zoology*, **7** (3) (Ed. Lankester). Black: London]

the primitively biramous appendages. They are characterised by a thin cuticle and the absense of hairs or setae. In the other four subclasses specialised respiratory structures are lacking.

A typical crustacean appendage consists of a basal proto-podite composed of a proximal coxopodite and a distal basi-podite (figure 2·5). These may both be further subdivided, the former into a precoxa and the coxopodite proper, and the basipodite into a probasipodite and a metabasipodite. Arising from the basipodite are the two rami of the appendage: an outer exopodite and an inner endopodite. The endopodite has five joints which, starting proximally, are called the ischio-podite, meropodite, carpopodite, propodite and dactylopodite. The exopodite and endopodite may each bear lateral pro-cesses, termed exites and endites respectively. The respiratory lobes or branchia are in many cases epipodites, and these arise from the coxopodite. The appendages themselves may be more or less cylindrical, as in the walking legs (pereiopods), or they may be flattened phyllopods. When the phyllopods are specifically involved in swimming they are referred to as swimmerets or pleopods.

Gills are present in the majority of branchiopods (the name means 'gill-feet'). In all cases they are simple plates formed from the epipodites of trunk appendages which are themselves flattened to form phyllopods.

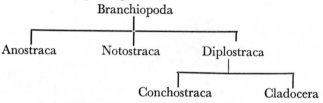

In the fairy shrimps (Anostraca) such as *Artemia* and *Polyartemia* (figure 2·6a) and the tadpole shrimps (Notostraca) such as *Apus* (figure 2·6b) there are between 11 and 19 pairs of trunk somites bearing phyllopods, all of which have branchial lobes. The remaining order (Diplostraca) is divided into the clam shrimps (Conchostraca) which possess 10 to 27 pairs of thoracic appendages, each with a gill blade (figure 2·6c), and the water fleas (Cladocera) with only four to six pairs of thoracic appendages. Most cladocerans (e.g. *Daphnia*) (figure 2·6d) have flattened gill-bearing appendages similar to those of the other branchiopods, but there are a few (Haplopoda) that have cylindrical rather than flattened appendages and these do not possess gills.

The literature concerning the presence of gills in the Ostracoda is scanty and somewhat conflicting. It would appear that most ostracods do not possess specialised respiratory structures, although it has been suggested that one of the lobes of the single

---

FIGURE 2·6 Gills of various branchiopod crustaceans. (a) 11th thoracic appendage of *Polyartemia forcipata* (Anostraca); (b) 7th thoracic appendage of *Apus crancriformis* (Notostraca); (c) 3rd thoracic appendage of *Estheria obliqua* (Conchostraca); and (d) 3rd thoracic appendage of *Daphnia pulex* (Cladocera). *end*, endite; *epi*, epipodite; *exo*, exopodite; *pro*, proepipodite. [All after Calman (1909) in *A Treatise on Zoology*, **7** (3) (Ed. Lankester). Black: London]

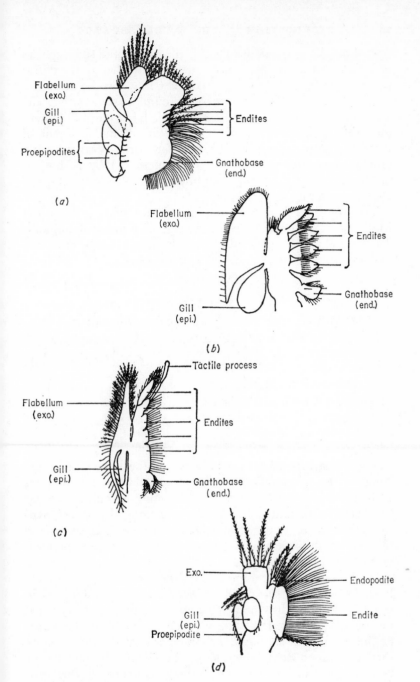

(a)

Flabellum (exo.)

Gill (epi.)

Proepipodites

Endites

Gnathobase (end.)

(b)

Flabellum (exo.)

Gill (epi.)

Endites

Gnathobase (end.)

(c)

Flabellum (exo.)

Gill (epi.)

Tactile process

Endites

Gnathobase (end.)

(d)

Exo.

Gill (epi.)

Proepipodite

Endopodite

Endite

B

pair of maxillae functions as a gill. In *Asterope* and some species of *Cypridina* there are plate-like structures attached to the dorsal posterior surface of the trunk and these may also be gills.

In the Cirripedia gills are only found in one order, the Thoracica (barnacles). *Lepas* for example has a pair of filamentous gills (epipodites) on the protopodite of the first pair of thoracic limbs. In addition the surface area of the mantle around the upper margin has been increased by folding to produce leaf-like projections, and these may also be involved in gaseous exchange.

The fourth subclass containing species with gills is the Malacostraca, an extremely large and varied group of animals. The branchiae are generally thoracic, but in some cases are abdominal.

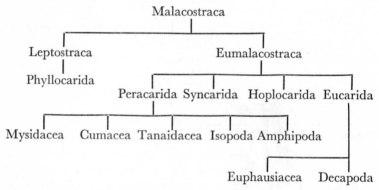

The simplest type of gill is that found in the Leptostraca (e.g. *Nebalia*) (figure 2·5). The thoracic legs are flattened phyllopods with well-developed epipodites functioning as gills, a similar arrangement to that found in the branchiopods. In the Syncarida the second to the eighth thoracic legs are pereiopods and these normally bear a pair of simple lamellar gills (epipodites) on the coxopodites.

Of the five orders comprising the Peracarida all except the Isopoda possess thoracic gills. These are probably simplest in the amphipods, the thoracic limbs of which are flattened pleopods. There are few gills and these are found on the coxopodites of the last six pairs of thoracic legs at the most. Again they are formed from epipodites, and are generally lamellar or vesicular, but in some cases branching occurs or the surface is ridged. In caprellids the number of gills is reduced to two pairs of simple lamellae on the fourth and fifth thoracic limbs, while in terrestrial species (beach fleas) gills are present but reduced in size. Accessory gills sometimes occur, and these may be on the thoracic appendages (some gammarids and cyamids) or on the first abdominal segments (talitrid beach fleas or sand hoppers).

Mysidaceans are similar in that the gills, when present (i.e. in lophogastrids), are also derived from thoracic epipodites, but are normally branched (figure 2·7a). *Mysis* itself does not possess gills and in the Mysidae the inner surface of the carapace acts as a site of oxygen uptake. The gills of euphausiaceans resemble those of the lophogastrids in that each thoracic leg bears an epipodite, usually branched, which acts as a gill (e.g. *Meganyctiphanes*) (figure 2·7c).

In all of the crustacean groups so far mentioned the gills are very simple, exposed and not covered by extensions of the carapace. However, in the other two peracaridan orders with thoracic gills, the Cumacea and the Tanaidacea, the gills are rather more complex (figure 2·7b), although they still do not reach the degree of complexity found among the decapods (p. 42). In the Cumacea a large epipodite on each first thoracic limb (first maxilliped) forms a series of filamentous gills which lie within, and form the floor of, a branchial chamber. The carapace is extended posteriorly to cover three or four thoracic segments, and anteriorly over the head to form a rostrum. The carapace and the sides of the body form the roof and walls of the branchial chamber. Similarly in the Tanaidacea it is only the epipodites of the first maxillipeds that form gills. The carapace covers the first two thoracic somites and encloses a small vascularised gill chamber on each side, the inner surface of the carapace also functioning as a site of oxygen uptake, as in the Mysidae. One of the simplest gills in decapods is that found in the carid shrimp *Pandalus* (figure 2·7d).

Two eumalacostracan groups possess gills associated with the abdominal appendages, which in both cases are flattened pleopods. These are the superorder Hoplocarida and one of the peracaridan orders, the Isopoda. In the former there are five pairs of pleopods, each of which bears a gill. The gill consists of an axial stem which arises from the anterior surface of the exopodite and bears a series of filaments.

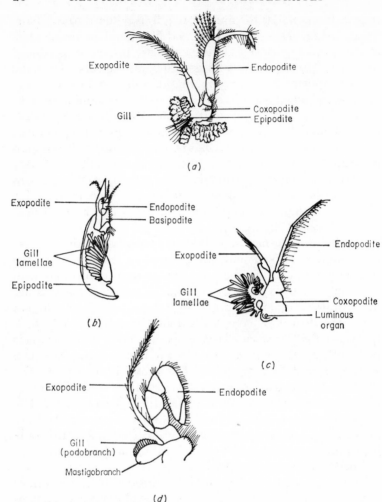

FIGURE 2·7 Appendages of various crustaceans. (*a*) 2nd thoracic appendages of *Gnathophausia longispina* (Mysidacea); (*b*) 1st maxilliped of *Diastylis stygia* (Cumacea); (*c*) 7th thoracic appendage of *Meganyctiphanes norvegica* (Euphausiacea); and (*d*) 2nd maxilliped of *Pandalus borealis* (Caridea). [All after Calman (1909), in *A Treatise on Zoology*, **7** (3) (Ed. Lankester). Black: London]

The Isopoda are a fascinating group including aquatic, semiterrestrial and fully terrestrial species. Typically there are six pairs of biramous abdominal appendages, both the expo-

dite and endopodite of which are modified for respiration.
There are, however, numerous variations on the basic plan.
They may be slight, as in the aquatic *Ligia*, where there are
only five pairs of limbs and in which the endopodites of the
second pair are modified in the male, not for respiration, but
as copulatory styli. In other genera the variations are more pro-
nounced. In some of the aquatic isopods which swim (rather
than crawl) the anterior limbs may be modified for swimming,
and only the posterior ones modified for respiration, or they
may all serve both functions. The gills are normally held flat
against the underside of the abdomen and are often protected
by an operculum. This may be formed from the first, second or
third pair of abdominal pleopods or, in the Valvifera, from the
uropods. As one may expect the more profound changes from
the basic plan are to be found in terrestrial species. In the
Oniscoidea, which contains both amphibious and terrestrial
species, the exopodite of each pleopod forms an operculum
which is folded over to protect the respiratory endopodite. In
the truly terrestrial isopods the operculum may contain a lung-
like cavity, as in the Oniscidae, or possess tube-like invaginations, as in the Porcellionidae and the Armadillidiidae. These
invaginations are called pseudotracheae. The gill is normally
a simple lamella, but it may be folded or possess projections,
particularly in some parasitic species.

Respiratory currents. In the branchiopods movements of the
trunk appendages generate water currents which ventilate the
gill surfaces and in addition act as feeding currents, and this
dual role is seen in many other crustaceans. In the Syncarida
the exopodites of the thoracic appendages generate the respiratory currents. This is taken a stage further in the Mysidacea
where the exopodite on each of the second to the eighth thoracic
appendages bears a slender setose flagellum to aid the circulation of water over the gills. This current is directed anteriorly
and is aided by movements of the maxillipeds. Similarly in the
Euphausiacea the thoracic exopodites are used to produce ventilatory currents, but in aquatic amphipods it is the more
anterior thoracic pleopods which produce the posteriorly
directed respiratory flow. Some amphipids are burrowers and
there is a tendency in these species for the antennae to be used
rather than the thoracic limbs.

The members of two peracaridan orders have rather complex respiratory systems. The Cumacea are small marine animals which live almost completely buried in sand or mud and they have become highly specialised for this mode of life. The carapace is inflated on each side of the body into a branchial chamber, the floor of which is provided by the filamentous gills themselves. It is also produced anteriorly in front of the head to enclose the large exopodites of the first pair of thoracic appendages (maxillipeds). These exopodites together with the rostrum form an exhalant siphon on each side of the head. Water is drawn in anteriorly over the mouthparts, circulates over the gills, passes upwards through the branchial chambers and is ultimately discharged through the exhalant siphons. Thus both the inhalant and exhalant opening are situated anteriorly, which is in keeping with the position which these fascinating little animals assume in their habitat. The water is pumped into the branchial chambers by movements of the epipodites bearing the gills, and possibly also by the second maxillae. In some species, such as *Diastylis*, this respiratory current is reinforced by movements of the endites of the first maxillipeds.

Tanaidaceans are mostly very small animals and are mainly marine. Like cumaceans they tend to live in mud and consequently show similar respiratory adaptations to this mode of life. The carapace covers and is fused to the first two thoracic segments, and encloses a gill chamber on each side of the body. The gills, which are formed from the epipodites of the first maxillipeds, lie in these chambers. In some species (e.g. *Apseudes*) ventilation is effected by movements of the gills themselves and of the exopodites of other thoracic appendages, which draw water into the branchial chambers under the posterior ventral margin of the carapace just in front of the bases of the third thoracic limbs. Thus the inhalant openings are quite posteriorly situated, rather than anteriorly as in the cumaceans, and this is possible because tanaidaceans only partly bury themselves. In some other genera, such as *Tanais*, the thoracic exopodites are absent and the branchial chambers less well developed. In addition to the branchial epipodites the inner surface of the carapace is vascularised and functions as a site of oxygen uptake. As in many other crustaceans the respiratory and feeding currents of both cumaceans and tanaidaceans are combined.

2.3. Insecta Several orders of insects contain species whose aquatic larvae or pupae possess gills. All insects which have any modifications for respiration possess a tracheal system, and this will be described in a subsequent chapter (chapter 4). Normally this system communicates directly with the exterior through openings called spiracles. However, in certain aquatic larvae there are no functional spiracles (the system is said to be 'closed') and respiration is effected by gills which contain fine branches of the tracheal system in much the same way that the gills of most other animals contain fine branches of the circulatory system. These gills are known as 'tracheal gills'. Some aquatic pupae possess 'spiracular gills' in which the peritreme and atrial region of one or more pairs of spiracles are drawn out to form long processes. Finally a number of insects possess so-called 'blood gills', which are normally tubular or vesicular projections containing circulatory fluid. However, in most cases these latter are not considered to be respiratory organs.

Altogether about a third of the insect orders include species which possess gills at some stage of their life cycle. Three of these—Ephemeroptera, Odonata and Pleoptera—belong to the Exopterygota; the others—Neuroptera, Trichoptera, Coleoptera, Diptera and Hymenoptera (only a few endoparasitic braconids)—to the Endopterygota.

Tracheal gills. Probably the simplest situation is encountered in the mayflies (Ephemeroptera), many of which possess seven pairs of simple lamellate gills borne laterally on consecutive abdominal segments and projecting from the sides of the body. All seven pairs may be single as in *Baëtis* or some may be double as in *Cloëon* (figure 2·8a, b). In others the gills are bifid with either simple (*Leptophlebia*) or multifurcate (*Habrophlebia*) branching (figure 2·8c, d). All of these larvae are reasonably good swimmers. In a number of the dorsoventrally flattened species which cling to stones, such as *Ecdyonurus*, some of the lamellate gills have basal tufts of branchial filaments (figure 2·8e). In *Ephemera*, which burrows in mud, the first pair are vestigial and the rest are biramous, each consisting of a pair of lamellae fringed with long filaments (figure 2·8f). These can be reflexed over the back and are normally held in this position.

Caenis has a similar habitat to that of *Ephemera* and possesses six pairs of gills (the first six). The first of these is reduced in size, but the two second gills are very large and quadrangular and cover the remaining pairs, presumably for protection. The resultant branchial chamber is guarded by fringes of setae which prevent particles of sand and mud from entering in the inhalant

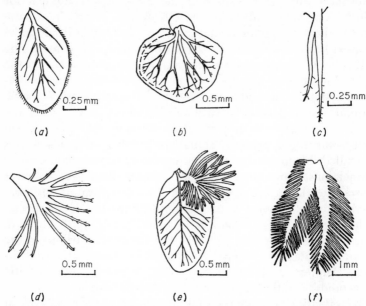

FIGURE 2·8 Tracheal gills of various mayflies (Ephemeroptera). All are of the 4th gill, except for that of *Leptophlebia* which is of the 3rd. (*a*) *Baëtis rhodeni*, (*b*) *Cloëon simile*, (*c*) *Leptophlebia cincta*, (*d*) *Habrophlebia fusca*, (*e*) *Ecdyonurus fluminum* and (*f*) *Ephemera vulgata*. [All after Eaton (1883–1888) *Trans. Linn. Soc. Lond.* (2), **3**]

current. In other mayflies the gills may be further reduced to five pairs, as in *Ephemerella* and *Prosopistoma*; the latter has an elaborate branchial chamber formed by the wing sheaths and the thoracic and abdominal terga. A few species, such as *Oligoneura* have, in addition to the abdominal gills, bunches of gill-filaments on various head and thoracic appendages.

The Odonata is divided into two suborders, the Anisoptera (dragonflies) and the Zygoptera (damselflies). The complexities of the respiratory system in the Anisoptera will be discussed later (p. 50), although the gills in many cases are simple lamellae. Most damselfly larvae possess three caudal gills: a median one derived from the dorsal appendage and two lateral ones, which are modified anal cerci. In young larvae the gills are always triangular in cross-section (triquetral). Normally they become swollen (saccoid) or flattened (lamellate), but sometimes retain their original shape, as in the case of the lateral gills of *Agrion*. A reduction of the gills and their tracheal system is found in the less fully aquatic species of *Melagrion*. A few species also possess lateral abdominal gills (e.g. *Anisopleura*).

Among the Neuroptera most megalopteran larvae possess lateral abdominal gills. *Corydalis* and *Chauliodes* for example bear a pair of unjointed or imperfectly jointed branchial filaments on each of the first eight abdominal segments, and *Corydalis* also possesses ventral spongy tufts of accessory gills. In *Sialis* filamentous lateral gills, composed of five segments, are found on the first seven segments and in addition there is a terminal filament of similar structure on the ninth segment. In the Planipennia (the second neuropteran suborder) only larvae of *Sisyra* possess gills.

In the other insect orders the gills are more variable in position. The stoneflies (Plecoptera) commonly possess them in the larval stages, and in the Eustheniidae there are five or six pairs of lateral abdominal appendages which function as gills, but these have disappeared in the other families in which gills, when present, are secondary and composed of tufts of filaments which are very variable in position. They sometimes persist into the adult stage, but when they do so are normally shrivelled and non-functional. However, the adult of *Pteronarcys* possesses thirteen pairs of apparently functional gill tufts on the ventral surface of the thoracic and the first two abdominal segments.

In the caddis flies (Trichoptera) they are again commonly present in the larval stages and frequently persist in the pupa (and in some genera are only present in the pupa). The gills are usually filamentous, and persist in a non-functional state in the adult of *Hydropsyche*.

In the Coleoptera and Diptera tracheal gills are rather more unusual. In the former they are always filamentous and only occur in members of a few families, members of the Gyrinidae for example possessing ten pairs of lateral abdominal gills fringed with hairs. In *Hygrobia* the gills are ventrally placed near the base of each pair of legs and on the first three abdominal segments, whereas *Peltodytes* has numerous elongated, jointed filaments on the dorsal surface of both thorax and abdomen. Their position is also very variable in the Diptera: thus in *Phalocrocera* they are segmental, arising from almost all segments, and are filamentous; alternatively they may be

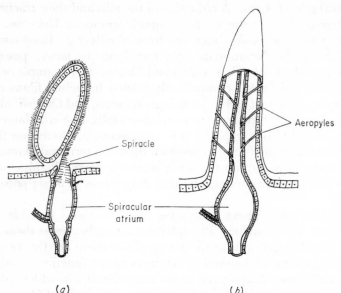

(*a*)   (*b*)

FIGURE 2·9 Two types of spiracular gills. (*a*) Simulid type with the atrial opening juxtaposed to the gill; (*b*) tipulid type, derived from a respiratory horn, with the atrium extending up into the gill. [After Hinton (1966) *Phil. Trans. Roy. Soc. B*, **251**]

ventral as in *Dicranota* (in which there are two pairs on the last abdominal segment), caudal as in the Culicidae or rectal as in *Simulium* and *Eristalis*.

Spiracular gills. These occur principally in the pupal stages of certain Diptera and Coleoptera, but are also found in the larvae of some of the Coleoptera, namely the Myxophaga. They are formed by modification either of the spiracle, of the surrounding area of the body wall, or of both, to form a projection (figure 2·9). The form of the gill varies considerably in different species and some examples are shown in figure 2·10. When the

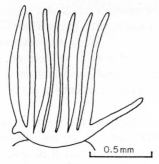

(*a*) Tipulidae: <u>Antocha bifida</u>

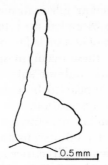

(*b*) Tipulidae: <u>Geranomyia unicolor</u>

(*c*) Tipulidae: <u>Dicranomyia trifilamentosa</u>

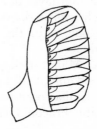

(*d*) Blepharoceridae: <u>Edwardsina tasmaniensis</u>

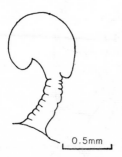

(*e*) Psephenidae: <u>Eubriinae</u>

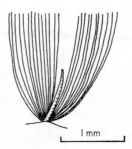

(*f*) Psephenidae: <u>Psephenoides gahani</u>

FIGURE 2·10 Spiracular gills of various pupae. (*a–d*) Diptera, Nematocera, (*e, f*) Coleoptera, Polyphaga. (For a transverse section of (*a*) see figure 4·11.) [*a* after Hinton (1966) *Proc. Roy. ent. Soc. Lond. A*, **41**; *b–d* after Hinton (1968) *Advances in Insect Physiol.*, **5**; *e* after Hinton (1966) *Phil. Trans. Roy. Soc. B*, **251**; *f* after Hinton (1947) *Proc. Roy. ent. Soc. Lond. A*, **22**] spiracular gill is formed from the spiracle itself the atrium is often extended up into it and connects with the surface via a series of channels (aeropyles) (figure 2·9). With very few exceptions these occur in species which inhabit a well-oxygenated aquatic habitat subject to rapid fluctuations in water level, and appear to be equally well-adapted for both aquatic and aerial respiration.

Similar structures, called respiratory horns, occur in some terrestrial dipterans and these are thought to have given rise to spiracular gills either by an increase in the number of aeropyles connecting the spiracular atrium with the surface (chapter 4) or by the development of structures on the surface capable of retaining a thin layer of air. Such structures constitute a plastron (which will be discussed in chapter 4) and all spiracular gills bear one, with the exception of some of those of the Chironomidae.

Blood gills. Many trichopteran larvae possess between four and six blood gills at the hind end of the abdomen, and they are found in a similar position in some tipulid (Diptera) larvae. In endoparasitic braconid (Hymenoptera) larvae a blood-filled

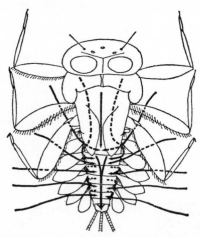

FIGURE 2·11 Dorsal view of the larva of *Ecdyonurus*, facing upstream. Arrows show the respiratory currents; in addition to those indicated there is a posterior flow of water under the animal which is directed up between the gills to the dorsal surface as indicated by the dotted arrows on the abdomen. [After Eastham (1937) *J. exp. Biol.*, **14**]

caudal vesicle is present. Apart from a few instances, like those just cited, the so-called blood gills are not held to be respiratory organs.

Respiratory currents. Many insect larvae rely on the flow of water over the body, produced either by movements of the animal itself or by water currents. However, many mayflies do produce proper respiratory currents (figure 2·11) and these are normally achieved by coordinated movements of the abdominal gills, although a few species have developed branchial chambers. The latter may be quite simple as in *Caenis*, where the second pair of gills is enlarged and covers the more posterior ones, or more sophisticated as in *Prosopistoma*, in which there are five pairs of abdominal gills enclosed in a well-developed chamber. The roof of this is a carapace, formed from the greatly developed prothoracic and mesothoracic terga which have fused with the sheaths of the mesothoracic wings. The side walls are formed by the metathoracic wing sheaths and the floor comprises the terga of the metathorax and first six abdominal segments. Water enters this branchial chamber through a pair of lateral apertures, flows over the gills and then passes out through a median opening. The pumping movements of dragonfly larvae will be considered in chapter 8.

2.4. Mollusca The complexities which the molluscan gill (ctenidium) attains in the Gastropoda, Bivalvia and Cephalopoda will be described later in the chapter (p. 51). However, the gill structure is relatively simple in three molluscan classes —Monoplacophora, Polyplacophora and Aplacophora. Gills are absent in the Scaphopoda (tusk shells).

Living representatives of the primitive Monoplacophora have only been discovered relatively recently and so far only three species have been described, all belonging to the genus *Neopilina* and all from deep-sea locations. The foot is almost circular and covers the bases of five or six pairs of serially arranged gills (figure 2·12). Each gill consists of a central axis from the ventral surface of which arise from two to eight flattened lamellae, the number depending partly on the species and partly on the length of the axis. There are, however, very short rudimentary lamellae on the dorsal side of the stem of at least the three anterior pairs of gills in *N. galatheae* and these alternate with the large ventral lamellae (figure 2·12).

In the chitons (Polyplacophora) a groove-like mantle cavity encircles the foot, and contained within this is a more-or-less complete row of gills on each side (figure 2·13). It is not certain whether there was an original single pair of gills or several pairs of segmental gills, but in any event there has been a secondary increase in their number in this class. As in the Monoplacophora each gill consists of a central axis from which arises a number of flattened lamellae which get shorter towards the tip. The central axis contains both afferent and efferent blood vessels and longitudinal muscles. Adjacent gills are held

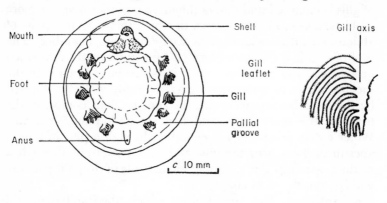

(*a*)   (*b*)

FIGURE 2·12 *Neopilina galathea*. (*a*) Ventral view to show the five pairs of gills in the mantle cavity; (*b*) A single gill with eight leaflets on one side of the gill axis and three or four rudimentary leaflets on the other. [After Lemche and Wingstrand (1959) *Galathea Rept.*, **3**]

closely together by long cilia on the anterior and posterior margins of the lamellae.

The members of two families in the Aplacophora possess respiratory projections in the rectal chamber. In the Chaeto-dermatidae there is a pair of horizontally oriented gills which are often visible externally; as in the previous two classes they are lamellate, with larger lamellae on the outer side of the axis. They bear cilia as well as sensory hairs and gland cells, and can be retracted. In the Neomeniidae there is a series of longi-tudinal gill folds—from a few up to about 40, depending on the species. They are normally simple, although in *Krup-pomenia minima* they are somewhat diverticulated.

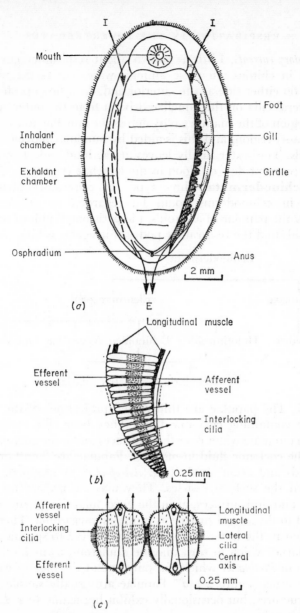

FIGURE 2·13 *Lepidochitona cinereus*. (*a*) Ventral view with gills and boundaries (.....) of shell plates drawn in the left pallial groove, and the division between inhalant and exhalant chambers (———) in the right pallial groove. (*b*) Posterior view of a single left gill. (*c*) Lateral view of a pair of gill filaments. → respiratory currents; ——→other currents. E, exhalant current; I, inhalant current. [After Yonge (1940) *Quart. J. micr. Sci.*, **81**]

Respiratory currents. Little is known about respiratory currents, except in chitons. In these animals water enters the mantle cavity on either side at the anterior end. As it flows posteriorly the lateral cilia on the lamellae drive it from the outer (inhalant) region of the mantle cavity across the lamellae to the inner (exhalant) region, and this is aided by muscular movements of the gills. The water finally leaves at the hind end. There is a countercurrent flow of blood in the lamellae (p. 12).

2.5. Echinodermata Three types of respiratory structure found in echinoderms could be regarded as simple gills; namely the papulae of asteroids, the peristomial gills of regular echinoids and the respiratory podia of irregular echinoids.

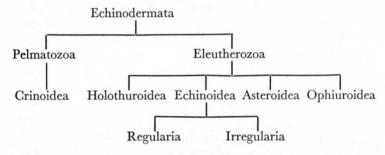

Papulae. The papulae are thin-walled projections of the body surface containing coelomic fluid. They bear cilia externally (which circulate water over their surface) and internally (which keep the coelomic fluid in motion). Papulae are found only in asteroids and occur singly (e.g. *Astropecten*) or grouped, lying between the skeletal ossicles. They may be present on both aboral and oral surfaces, as in the Forcipulata (e.g. *Asterias*) or limited to the aboral surface as in the Phanerozonia. Their distribution in the latter may be further limited to specific areas (papularia) which in turn may be numerous, as in *Linckia*, or few, as in *Pectinaster* which has only one at the base of each of its five arms (figure 2·14*a*). Papulae are usually simple tube-like structures, but occasionally exhibit branching (e.g. *Luidia*) (figure 2·14*b*).

Peristomial gills. Peristomial gills are present only in the regular echinoids. Like papulae they are thin-walled evaginations of the body surface which bear cilia externally and internally, but unlike them they are always branched (figure 2·14*d*). They are

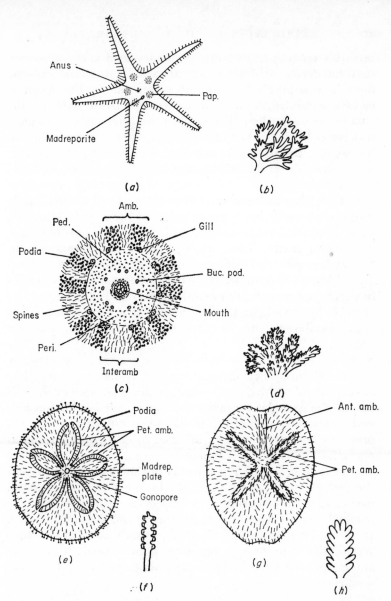

FIGURE 2·14 (a) Aboral view of *Pectinaster* to show papulae grouped into papularia; (b) papula of *Luidia*; (c) oral view of *Tripneustes ventricosus* to show gills; (d) gill of *Tripneustes*; (e) aboral view of *Clypeaster roseaceus* to show the five petaloid ambulacra typical of clypeastroids; (f) branchial podium of *C. roseaceus*; (g) aboral view of *Meoma ventricosa*, showing the unmodified anterior ambulacrum of spatangoids; (h) branchial podium of *Spatangus*. [All after Hyman (1955) *The Invertebrates*, Vol. IV. McGraw Hill: New York]

in direct continuity with the peripharyngeal coelomic cavity that surrounds Aristotle's Lantern and so contain coelomic fluid. This is pumped in and out of the gills by the alternate elevation and depression of the lantern and by the action of the cilia lining the coelom. The cilia on the outer surface circulate water over the gills. When present the gills lie in a circle around the edge of the peristome and they number ten, two in each interradius (i.e. at the beginning of each interambulacral area) (figure 2·14c).

In the Cidaroidea and the Echinothuriidae (the latter family contains all the living representatives of the Lepidocentroida) pouches of the lantern coelom project upwards into the main perivisceral cavity. These are referred to as internal gills or Stewart's organs and their presence tends to preclude external gills. Thus the Cidaroidea is the only order of regular echinoids in which peristomial gills are always lacking. In some members of the Echinothuriidae they are absent or at least rudimentary (e.g. *Tromikosoma*), but in others such as *Calveriosoma* they are present.

Respiratory podia. Podia or 'tube-feet' are present in the majority of echinoderms. Their original function is uncertain, but they have probably always subserved respiration. However, it is only in the irregular echinoids that some have become specialised respiratory structures; in all other groups they are all primarily concerned with locomotion. Most irregular echinoids normally live partly or completely buried in sand. There are two main groups, the Clypeastroida (includes sand dollars) and the Spatangoida (heart urchins). In the former the aboral part of each ambulacrum is expanded into a petal shape (the petaloid) which bears the respiratory podia (figure 2·14e, f). In the heart urchins only four of the ambulacra are so converted, the anterior one retaining normal podia which the animal uses to keep the respiratory passage to the surface of the sand clear (figure 2·14g, h).

COMPLEX GILLS

For convenience in dealing with gills from a comparative standpoint, and bearing in mind the inherent problems of classification mentioned at the start of the chapter, only those

with a fairly simple structure have been described so far. The
more complex types will now be dealt with, i.e. those of merostomes, dragonflies and higher crustaceans and molluscs. Gills
of great complexity are not found in annelids or echinoderms.
2.6. Merostomata This arthroped class can be divided into
the extinct Eurypterida and the Xiphosura. The body of
merostomes is divided into an anterior prosoma and a posterior
opisthosoma, and in both subclasses it is the opisthosomatic
appendages that form the gills. In the Xiphosura, of which the

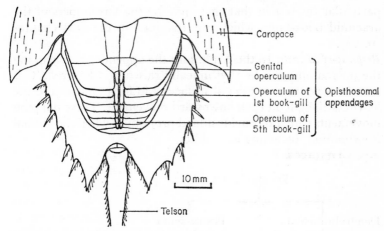

FIGURE 2·15 Opisthosoma of Limulus from the ventral surface

living representatives are the so-called 'horseshoe crabs' or
'king crabs', the paired appendages of the second to the sixth
opisthosomal segments are involved in respiration. Each consists of a two-jointed endopodite and an elongate exopodite.
The exopodites of each pair are joined in the ventral mid-line
to form a membranous flap (figure 2·15). On the posterior
surface of each flap there is a series of from 100 to 200 oval
gills with stiffened outer margins, forming a structure known as
a book-gill. The first pair of opisthosomatic appendages have
also fused with one another in the mid-line and form a genital
operculum which covers, and thus presumably protects, the
book-gills from damage (figure 2·15). The last pair of appendages possess an external, spatulate process which is inserted
under the operculum and cleans the gills.

The respiratory structures of the eurypterids, which existed from Cambrian to Permian times, also appeared to be book-gills, in so far as function can be determined from fossil evidence and from the existence of apparently homologous structures in the Xiphosura. They differ from those of the Xiphosura in two ways: firstly the operculum, formed from the first pair of opisthosomatic appendages, also appears to have possessed gills; secondly the appendages bearing the gills on the second to fifth opisthosomatic segments were not joined in the mid-line.

These external lamelliform gills of the merostomes are of particular interest in that they may be the forerunners of the arachnid book-lungs which are described in the next chapter (p. 66).

Respiratory currents. In the Xiphosura the appendages (including the operculum) can be rhythmically raised and lowered. This circulates water over the gill surfaces and forces blood into the leaves of the gill on each downward movement. These flapping movements can also provide added propulsion when the animal is swimming, by acting as paddles.

2.7. Crustacea

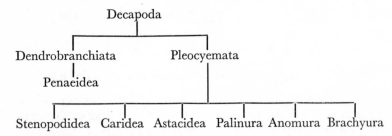

In decapod crustaceans the gills (figure 2·16) are attached either to the thoracic appendages or close to their bases, and are enclosed in a branchial chamber (figure 2·17) formed by lateral extensions of the carapace called the branchiostergites. There are up to four pairs of gills associated with each thoracic segment and thus, since the number of thoracic segments is eight, a total of 32 gills is possible on each side of the body. However, no known decapod has this number and there are normally far less. As far as is known the actual maximum is 24, as occurs in the penaeid *Benthesicymus*. At the other end of the scale the pea crab *Pinnotheres* has only three, while in between these two extremes lobsters and crayfish (Astacidea) possess around 20 and marine brachyuran crabs tend towards eight or nine.

Of the four possible gills associated with each half-segment,

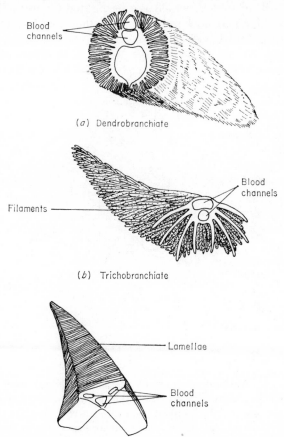

(a) Dendrobranchiate

(b) Trichobranchiate

(c) Phyllobranchiate

FIGURE 2·16 The three types of gill found in decapod crustaceans. [After Calman (1909) in *A Treatise in Zoology* (Ed. Lankester) and after Meglitsch (1967) *Invertebrate Zoology*]

one may be attached to the body wall above the base of the appendage and is called a pleurobranchia (in actual fact it is attached to the precoxa, but this has become incorporated into the body wall); two are attached to the articulating membrane between the body wall and the coxopodite (arthrobranchiae), and the fourth is attached to the coxopodite (podobranchia) and is formed from part of the epipodite.

Each gill consists of a central axis containing afferent and efferent blood vessels, which are perforated along their course to allow blood to flow in and out of the lamellae. However, most of the blood is directly conveyed from the afferent to the

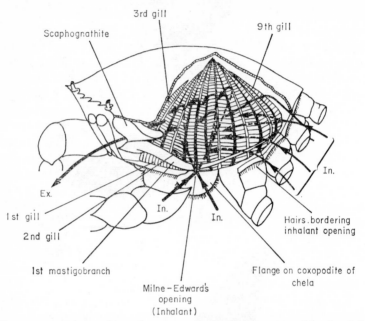

FIGURE 2·17 Anterolateral aspect of *Carcinus maenas* with the left branchial chamber exposed to show the gills and ventilatory currents. Inside the branchial chamber the dotted arrows indicate currents in the hypobranchial space; the solid arrows currents in the epibranchial space. *Ex.*, exhalant current; *In.*, inhalant current. [After Hughes, Knights and Scammell (1969) *J. exp. Biol.*, **51**]

efferent vessel round the periphery of each lamella in the marginal canal. Below the cuticle of the lamella there are two layers of epithelial cells and these become drawn out at certain points to form pillar junction cells. There are three basic gill types, differing from one another in the form of the filaments (figure 2·16):

Dendrobranchiate: two series of primary filaments. Each

primary filament is complexly branched to form a gill of many tiny filaments (only in the Penaeidea).

Trichobranchiate: a series of unbranched filaments, arranged in various patterns around the axis (e.g. crayfishes).

Phyllobranchiate: plate-like filaments, usually arranged in two series (e.g. *Carcinus*, *Leander*).

All the gills in any one species are of the same type, but, as far as the Pleocyemata are concerned, there is no real constancy of structure within the various suborders. Thus, although most anomurans possess phyllobranchiate gills some have trichobranchiate ones and others have gills intermediate between the two. Phyllobranchiate gills are also present in most Caridea and Brachyura. Trichobranchiate is the predominate gill type in the Astacidea but also occurs in some anomurans and a few primitive brachyurans. In addition to this complexity there is a range of structure within each type and this is well illustrated by the phyllobranchiate gill. In a brachyuran such as the common shore crab *Carcinus* it is attached ventrally and the lamellae arise in pairs on either side of the central axis; but in an anomuran such as *Porcellana* the gill is attached laterally and has two branches, a short ventral one and a longer dorsal one, and the central axis is more circular in cross-section. Again the lamellae arise in pairs, but are elongate in *Porcellana* as against almost quadrangular in *Carcinus*. In the Caridea the arrangement is similar to that found in *Porcellana*.

The shape of the carapace is variable and it may or may not cover all of the thoracic segments dorsally. However, the branchial chambers are always present and basically similar in form in the different groups. In *Carcinus* for example the carapace is extended laterally, almost in the horizontal plane, and then curves ventrally and inwards to fit closely against the thorax, the exopodites of the maxillipeds and the bases of the claw and legs; but several openings are left, namely:

(a) Small slits, one above each leg (Inhalant).

(b) A medium opening above the coxopodite of the claw (Milne-Edwards opening) (Inhalant).

(c) A large opening in front of the mouth (Exhalant).

The lateral wall of the thorax has become inclined to an angle of about 45° to the vertical and the gills thus lie at a similar angle.

At this juncture I must mention two other structures: the scaphognathites, which beat rhythmically to generate the respiratory current, and the mastigobranchs, which clean the gills. The scaphognathites (sometimes called gill-bailers) are plate-like structures formed from the enlarged exopodites of the maxillae, possibly also fused with the epipodites (figure

TABLE 2.1. The gill arrangements in (a) *Cancer pagurus*, (b) *Eupagurus bernhardus* and (c) *Galathea squamifera*

| | Thoracic appendages | | | | | | | | |
	1	2	3	4	5	6	7	8	Total
(a) *Cancer pagurus*									
Podobranchiae	—	1	1	—	—	—	—	—	2
Anterior Arthrobranchiae	—	1	1	1	—	—	—	—	3
Posterior Arthrobranchiae	—	—	1	1	—	—	—	—	2
Pleurobranchiae	—	—	—	—	1	1	—	—	2
Epipodites	(1)	(1)	(1)	—	—	—	—	—	(3)
Total	(1)	2+(1)	3+(1)	2	1	1	—	—	9+(3)
(b) *Eupagurus bernhardus*									
Podobranchiae	—	—	—	—	—	—	—	—	—
Anterior Arthrobranchiae	—	—	1	1	1	1	1	—	5
Posterior Arthrobranchiae	—	—	1	1	1	1	1	2	7
Pleurobranchiae	—	—	—	—	—	—	1	—	1
Epipodites	—	—	—	—	—	—	—	—	—
Total	—	—	2	2	2	2	3	2	13
(c) *Galathea squamifera*									
Podobranchiae	—	—	—	—	—	—	—	—	—
Anterior Arthrobranchiae	—	—	1	1	1	1	1	—	5
Posterior Arthrobranchiae	—	—	1	1	1	1	1	—	5
Pleurobranchiae	—	—	—	—	1	1	1	1	4
Epipodites*	(1)	—	(1)	(1)	(1)	(1)	—	—	(5)
Total	(1)	—	2+(1)	2+(1)	3+(1)	3+(1)	3	1	14+(5)

(a) from Pearson (1908) *L.M.B.C. Memoirs,* **16.** (b) from Jackson (1913) *L.M.B.C. Memoirs,* **21.** (c) from Pike (1947) *L.M.B.C. Memoirs,* **34**
* In *G. strigosa* there is only one epipodite, and this on the third maxilliped

2·17). The mastigobranchs (gill-rakers) are borne on the epipodites and are long, narrow processes (figure 2·17). In *Carcinus*
there are three mastigobranchs on each side; that of the first
maxilliped lies external to the gills, whereas the others (on the
second and third maxillipeds) are internal. In some astacidians
such as the crayfishes the podobranchs are fused with the mastigobranchs, but in others (e.g. *Nephrops*) they are as free as in
Carcinus. The number and pattern of the gills and mastigobranchs is perhaps most clearly presented in the form of a
table (table 2.1).

Those crabs which have become amphibious tend to have
few gills and the surface area of these is generally reduced,
trends which are continued further in the land crabs (table
2.2). There are four groups of land crabs; the gecarcinids, some

TABLE 2.2. Gill surface area in aquatic, amphibious and
terrestrial crabs

	Gill surface area (mm²/g)	Habitat
Callinectes	1367	Aquatic (active)
Libinia	748	Aquatic (sluggish)
Cancer pagurus	425	
Carcinus maenas	777	Intertidal (active)
Uca	624	Intertidal
Ocypode	325	Terrestrial

Data from Gray (1957) *Biol. Bull.*, **112,** and Scammell (1971) Ph.D.
Thesis (Bristol)

grapsids and some potamonids (all belonging to the Brachyura),
and the coenobitids (Anomura). They all retain some gills, but
in coenobitids these are supplemented in their role by vascularisation of the walls of the branchial chambers (p. 70).

Respiratory currents. The branchial chamber of aquatic brachyurans is divided by the gills into a dorsal epibranchial (exhalant)
and a ventral hypobranchial (inhalant) chamber. In *Carcinus*
the first two gills are small, but the other seven curve dorso-
medially from the bases of the legs to effect this separation of
the two chambers. The epibranchial chamber narrows anteriorly to form an exhalant passage which terminates at the large

opening in front of the mouth. The scaphognathite lies in this exhalant passage and its movements drive water out of the opening. Water is thus drawn in through the other apertures (figure 2·17). Most of the water enters the hypobranchial chamber via the opening above the claw, but some also enters via the small openings above the four walking legs. It flows between the gill lamellae into the epibranchial space and is then drawn anteriorly towards the exhalant passage. There is a process on the coxopodite of the third maxilliped which can be used to close the anterior part of the aperture above the claw when required. When this aperture is fully open the water entering it irrigates most of the gills, but when the anterior part is closed only the first five receive any water via this route. The irrigation of the gills is shown in figure 2·17. In *Carcinus* the ventilatory activity shows a tidal rhythm with a maximum occurring at, and just after, the time of high tide. While normal ventilation is occurring the beating of the scaphognathite produces rhythmic changes in pressure in the branchial chamber (3–12 mm water) generally superimposed on a maintained pressure of up to 10 mm water below that of the surrounding water.

In the lobsters and crayfish the branchial chambers are more rounded and less completely enclosed, whereas the anomurans show an intermediate condition. In the other orders the ventral margins of the carapace fit loosely against the sides of the body and so water can enter the branchial chambers anywhere along the ventral and posterior margins of the carapace.

The respiratory current in *Carcinus* is periodically reversed and the main reason for this is probably to clear the hypobranchial chamber of particles after these have been dislodged from the ventral surface of the gills by the action of the mastigobranch of the third maxilliped. (Particles cleared from the dorsal gill surfaces by the mastigobranch of the first maxilliped are in the epibranchial chamber and so will be carried away during forward ventilation.) In addition periodic reversals would disturb any stagnant diffusion barrier which was tending to form and would allow the animal to aerate its gill chambers when it is partially buried in sand or incompletely immersed in water. These reversals last up to five seconds and produce a pressure in the epibranchial chamber of 0–22 mm water above that of the surrounding water. Reversals also occur in the squat lobster (*Galathea*), but here they are irregular and weak. In animals which burrow the current tends to be reversed when they bury themselves, while in *Porcellana* the respiratory current is usually maintained by one side only while the other rests.

In those animals which live in silty or muddy conditions various degrees of protection of the branchial chambers are exhibited. *Galathea* for instance has simple setae projecting over the inhalant apertures and this is taken a step further in *Porcellana* in which the setae are bipinnate and form a thick curtain over these apertures. It is interesting to note that, having gone to this length to protect its gills, *Porcellana* has no mastigobranchs (in *Galathea* the number is variable and reduced to one in *G. strigosa*). Both *Porcellana* and *Galathea* use their fifth pereiopods (which are very thin and chelate) to remove sediment from the branchial chambers. In some species the antennules are held together to form an inhalant siphon, whereas in others the antennal scales (*Metapenaeus*) or the antennae (*Corystes cassivelaunus*) are similarly employed. In some cases exhalant siphons are formed by the third maxillipeds (e.g. *Calappa*).

In most semiterrestrial decapods the branchial chambers are partially filled with water, which may be retained by a flap on each maxilliped and aerated by movements of the scaphognathites which circulate air through the branchial chambers (e.g. *Ocypode* and *Grapsus*) (p. 70). A different method of aeration is seen in *Eriocheir* for example, where the water is pumped out in the usual way and then flows in channels back over the body to the inhalant openings at the base of the legs.

Truly terrestrial crabs do not carry water in their branchial chambers, although gills are still retained. In the gecarcinids and grapsids (e.g. *Geograpsus*) the lamellae of the gills are fairly rigid and held apart to allow air to circulate between them, and their surfaces are kept moist. The scaphognathite still retains its role of producing the respiratory current, but is often aided by up-and-down movements of the carapace margins (p. 70). In the gecarcinids, at least *Cardisoma guanhumi* and *Gecarcinus lateralis* maintain a high level of humidity in the branchial chambers by a pair of pericardial sacs, which bulge into the branchial chambers and can absorb and store water.

2.8. Insecta The most complex gill system in this group is that found in the larvae of the dragonflies (Odonata, Anisoptera). The gills are completely enclosed in a modified region of the hind-gut called the branchial chamber. There are six longitudinal folds in the branchial chamber and the gills are borne either directly on these, or indirectly via an associated double series of cross-folds. The shape and the attachment of the gills are variable and serve as a basis for their classification (figure 2·18):

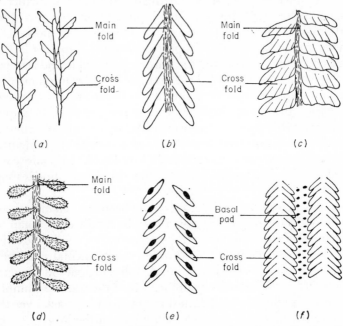

FIGURE 2·18 Examples of different types of tracheal gills found in the branchial chamber of the dragonfly larva. (*a*) Simplex, undulate (*Austrogomphus*); (*b*) duplex, implicate (*Austroaeshna*); (*c*) duplex, foliate (*Aeshna*); (*d*) duplex, papillofoliate (*Anax*); (*e*) duplex, archilamellate (*Synthemis*); (*f*) duplex, neolamellate (*Diplacodes*). [After Tillyard (1917) *The Biology of Dragonflies*. Cambridge University Press: Cambridge, England]

> Simplex: longitudinal folds bearing gills; supported on either side by cross-folds.
> Undulate: free edge of each gill undulated.
> Papillate: gills divided up into elongate filaments.

Duplex: longitudinal folds non-functional or absent. Gills attached to cross-folds.

Implicate: obliquely placed, concave tile-like gills.

Foliate: basally-constricted leaf-like gills.

Papillofoliate: basally constricted papillate gills.

Lamellate: broad-based, projecting gills. Flat and plate-like.

> Archilamellate: twelve lamellae in each row. Large basal pads.

> Neolamellate: numerous lamellae in each row. Small basal pads.

There are three pairs of longitudinal tracheal trunks (chapter 4), two of which (dorsal and visceral) send branches to the gills. The third pair (lateral) join the visceral towards the hind end of the branchial chamber. The initial branches from the longitudinal trunks are termed primary efferent tracheae and these divide, first to form secondary efferent tracheae, and then in the gills into numerous tracheoles. Each tracheole forms a complete loop within the gill and is connected at each end to the same secondary efferent trachea. At the base of each gill there is a pad of cells with which is associated some adipose-like tissue. In addition to the tracheoles, the gill also contains pigment granules.

Posterior to the branchial chamber, and separated from it by a valve, there is a small muscular chamber called the vestibule. There are also valves controlling the opening from the branchial chamber into the more anterior part of the gut, and the opening (anus) of the vestibule to the exterior.

Respiratory current. The respiratory current is tidal, water passing out of the anus during expiration and in again during inspiration. This is caused by muscular movements and the precise nature and control of this process is described in some detail in chapter 8.

2.9. Mollusca Of the three molluscan classes which will be dealt with here—Gastropoda, Bivalvia and Cephalopoda—the cephalopods have the least variable gills (ctenidia). Two of the cephalopod subclasses contain living representatives, namely the Nautiloidea, which possess two pairs of ctenidia (or at least the living representative *Nautilus* does) and the Coleoida (including octopods and squids), in which a single pair of ctenidia is a characteristic feature (for example, *Sepia*, figure 2·19). In *Nautilus* the ctenidia are only attached basally, but in

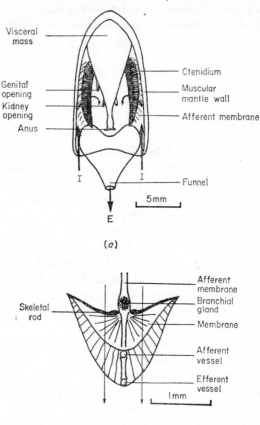

(*a*)

(*b*)

FIGURE 2·19 *Sepia*. (*a*) Mantle cavity from the ventral (morphologically posterior) surface. (*b*) Section through a ctenidium to show the ctenidial filaments. Arrows indicate respiratory currents. [After Yonge (1947) *Phil. Trans. Roy. Soc. B*, **232**]

the Coleoida there is in addition an afferent supporting membrane. In both groups flattened gill filaments are arranged alternately on opposite sides of a central axis which contains the main afferent and efferent blood vessels. The surface area of the filaments is increased by folding and, in the Coleoida, they contain beds of capillaries. However, unlike in the gastropods and bivalves they do not bear any cilia.

The gastropods are an intriguing group of animals. Primitively they are thought to have possessed a single pair of ctenidia lying in the mantle cavity and pointing posteriorly. However,

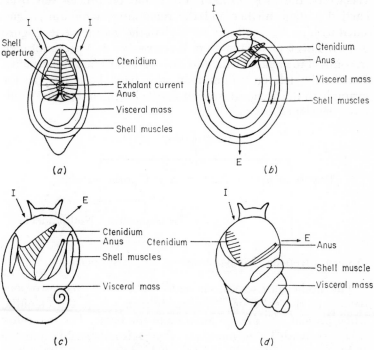

FIGURE 2·20 Ctenidia in Gastropoda, Prosobranchia. (*a*) Archeogastropoda, *Fissurella*; (*b*) Archeogastropoda, *Acmaea*; (*c*) Mesogastropoda; (*d*) Neogastropoda. *a–c* are aspidobranchs and *d* is a pectinibranch. Arrows indicate the respiratory currents. E, expiratory; I, inspiratory. [After Yonge (1947) *Phil. Trans. Roy. Soc. B*, **232**]

the living members of this class have all undergone a process called torsion, which has involved some of the organs, including the ctenidia, in an anticlockwise rotation through 180° with respect to the head and foot. In some archeogastropods such as *Pleurotomaria* and *Fissurella* there are still two anteriorly directed ctenidia (figure 2·20*a*). However, in most cases the right ctenidium has been lost and the left one has come to lie partly on the right side of the body, as for example in the archeogastropods *Trochus* and *Acmaea* (figure 2·20*b*) and in the mesogastropods (figure 2·20*c*). (In referring to left and right ctenidia the accepted convention is to take the torted condition as the basal one, i.e. primitive left becomes right and vice-versa.) In the neogastropods the reduction process has been carried a stage further and the remaining ctenidium is one-sided with its axis built into the wall of the mantle cavity (figure 2·20*d*). This type of ctenidium is referred to as a pectinibranch in contrast to ctenidia with two rows of leaflets (aspidobranchs). Finally the left ctenidium may also have been lost, as for example in the archeogastropod *Patella* in which secondary gills have developed (p. 54).

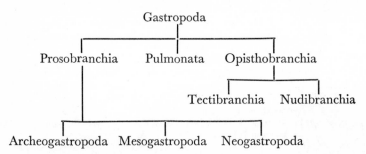

The archeogastropods, mesogastropods and neogastropods belong to the subclass Prosobranchia in which the ctenidia, when present, are always anterior to the heart. The two other subclasses with living members, the Opisthobranchia and the Pulmonata, are thought to have evolved from primitive proso-branchs. The opisthobranchs display various degrees of detor-sion and are divided into two groups, the tectibranchs and the nudibranchs. The former normally possess a single ctenidium (e.g. *Aplysia*), but in some cases this has been lost. On the other hand mantle cavity, ctenidia and shell have all been lost in the nudibranchs and only secondary gills are present. These are arranged either around the anus (e.g. *Archidoris*, figure 2·21*a*), in rows on the dorsal surface (when they are called cerata) (e.g. *Coryphella*, figure 2·21*b*) or in rows laterally under the edge of the mantle. The cerata may be club-shaped (*Coryphella*) or branched in various ways (*Idulia* and *Dendronotus*) (figure 2·21*d–f*).

The pulmonates are devoid of ctenidia and, as may be deduced from the name, the wall of the mantle cavity is vascularised. This has occurred as an adaptation to a terrestrial environment. However, a number of pulmonates have returned

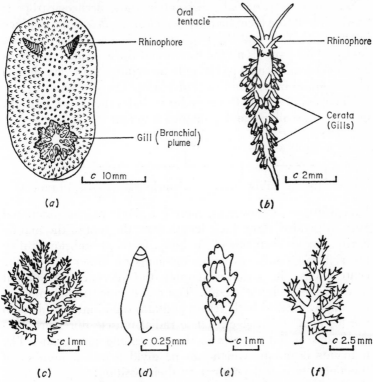

FIGURE 2·21 Dorsal views of (a) *Archidoris* (*Doris*) *flammea* and (b) *Coryphella pedata* (*Eolis landsburgi*). (c) Gill from branchial plume of *Archidoris* (*Doris*) *flammea*. (d–f) Cerata from *Coryphella pedata* (*Eolis landsburgi*), *Idulia* (*Doto*) *coronata* and *Dendronotus frondosus* (*D. arborescens*). [After Alder and Hancock (1845–1855) *A Monograph of the British Nudibranchiate Mollusca*. Ray Society: London]

to an aquatic environment and some of these have become completely re-adapted, in that they can obtain their oxygen requirements from the water and do not have to surface for air. Of these the fresh-water limpets (Ancylidae and Planorbidae) and the members of two marine families (Amphibolidae and Siphonariidae) have developed secondary gills. In the marine

c

species these are contained within the mantle cavity, but in the fresh-water limpets the mantle cavity is reduced and the single secondary gill (pseudobranch) is a conical projection from the left side of the foot.

Secondary gills are also present in those archeogastropods in which both the ctenidia have been lost. In *Patella* for instance between the foot and the mantle there is a secondary mantle cavity which extends all the way round the animal. It contains a series of folds termed pallial gills, an arrangement very similar to that found in the chitons (Polyplacophora) (p. 36).

The class Bivalvia (Pelecypoda) includes the clams, mussels, oysters and scallops, and is divided into four subclasses:

Protobranchia Filibranchia Eulamellibranchia Septibranchia

Except in the septibranchs, there is a single pair of functional gills, suspended along their length from the roof of the mantle cavity, one on each side of the body. In the protobranchs they are simple ctenidia with flat filaments on either side of the central axis, the latter containing the afferent and efferent blood vessels (figure 2·22a). There are longitudinal muscles running ventral to both vessels and also radial muscle fibres in the filaments themselves. These radial muscle fibres are continuations of the muscles in the suspending membrane. The filaments bear cilia which can be divided into a number of functional groups dependent on their position:

(a) Lateral. These drive water up over the filaments and aid in the production of the respiratory currents. In the filibranchs and eulamellibranchs they are solely responsible for this.

(b) Laterofrontal. These are poorly developed in protobranchs, but in filibranchs and eulamellibranchs are very well developed and serve to deflect food particles onto the frontal surface of the filaments.

(c) Frontal. These beat inwards to the mid-line between the ctenidia, where there is an anteriorly directed current. In some (e.g. *Lembulus*) this inward current is produced by the outermost frontal cilia, the inner ones beating towards the axial

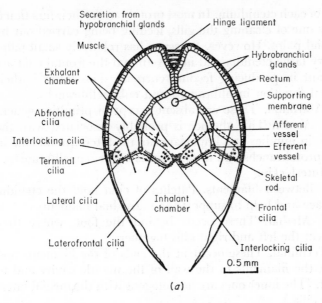

(a)

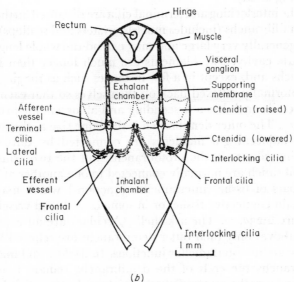

(b)

FIGURE 2·22 Protobranchia. Transverse sections of (a) *Nucula turgida* and (b) *Malletia obtusata* to show the form and position of the ctenidia. [After Yonge (1941) *Phil. Trans. Roy. Soc. B*, **230**]

groove of each ctenidium. In most protobranchs their function is entirely one of cleaning the gills, feeding being carried out by the labial palps. However, *Solenomya* has relatively small palps and very large ctenidia and in this animal the frontal cilia are important in producing feeding currents. This latter is their principal function in filibranchs and eulamellibranchs.

(d) Abfrontal. These are absent except in primitive genera such as *Nucula*. This absence is probably associated with the increase in number and importance of the laterofrontal cilia in higher protobranchs and in filibranchs and eulamellibranchs.

(e) Interlocking

(i) Between filaments. Patches of cilia hold the ctenidia in place and help to support the individual filaments.

(ii) Medial. These occur behind the foot, where they connect the left and right gills together.

(f) Terminal. These occur at the ends of the filaments and connect the filaments to the wall of the mantle cavity and to the foot. (The inner ones are homologous with the medial interlocking cilia.)

The interlocking and terminal cilia are discussed further below.

In filibranchs (includes mussels, oysters and scallops) the gills are generally very large and may extend the whole length of the mantle cavity. The filaments are much longer than in protobranchs and, except in a few instances such as the jingle shells, are normally reflexed back on themselves so that each half-gill or demibranch has a descending and an ascending arm (figure 2·23a). The outer demibranch is the same size as the inner one. The adjacent gill filaments are connected by tufts of interlocking cilia as in the protobranchs, but the two arms of each demibranch are normally connected to one another at intervals by bars of tissue (interlamellar junctions) which may simply contain connective tissue, or in some cases blood vessels as well (figure 2·23a, c). The ark shells (Arcidae) are an exception to the above: only their outer demibranchs are reflexed back and there are no interlamellar junctions. In *Mytilus* and many other filibranchs the ends of the demibranchs remain free, but in others (e.g. the oyster, *Ostrea*) they have become fused with the mantle wall and the base of the foot at their outer and inner ends respectively. In primitive filibranchs (e.g. *Pecten*) the efferent vessel is still in the central axis along with the afferent vessel, but in many it has become divided into two vessels, one of which runs along the outer end of each demibranch.

In the eulamellibranchs (clams, cockles, fresh-water mussels, etc.) the gill structure is even more complex. The basic pattern

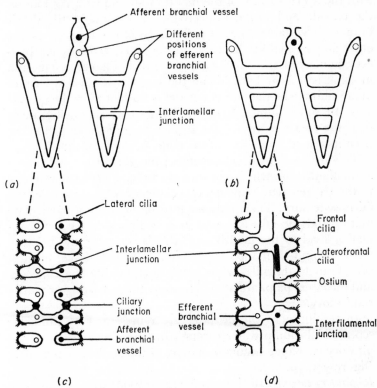

FIGURE 2·23 Transverse sections to show the reflexed nature of the ctenidial filaments of (*a*) a filibranch and (*b*) a eulamellibranch with the corresponding frontal sections (*c*, *d*). [(*a*, *b*) after Yonge (1947) *Phil. Trans. Roy. Soc. B*, **232**. (*c*, *d*) after Barnes (1968) *Invertebrate Zoology* Saunders: Philadelphia]

of reflexed demibranchs is present, but adjacent filaments are now no longer connected to one another simply by tufts of interlocking cilia, but rather by bars of tissue (interfilamental junctions) which are always vascular (figure 2·23*b*, *d*). Also there are many more interlamellar junctions, again always vascular. The net result of this is to form a fairly rigid gill which contains vertical water tubes within the demibranchs, and which connects with the mantle cavity via openings called ostia. The frontal edges of the filaments are not involved in this fusion. As in *Ostrea* the ends of the ascending limbs are fused with the upper surface of the mantle wall and with the foot on the outside and inside respectively. The form of the demibranchs is variable: they may both be of similar size, or the outer may be slightly shorter than the inner, as in *Unio*. However, in most cases the outer is much shorter than the inner, and in some it is bent back above the latter (e.g. *Lyonsia*). As in many filibranchs the efferent vessel is no longer in the central axis. In some filibranchs (e.g. *Ostrea*) and eulamellibranchs (e.g. the cockle, *Cardium*) the surface area of the lamellae has been increased by folding of the edges of the filaments.

This tremendous development of the gills in the filibranchs and eulamellibranchs is not thought to have been entirely due to the oxygen requirements, especially since many species are fairly sedentary. These animals feed on very small particulate matter and the gills are used as a straining device for channelling food to the mouth, and this is probably the primary factor behind their increase in size. As stated above, in most protobranchs the gills do not reach anything like the same size and the labial palps are used in feeding, the gills being primarily respiratory.

In septibranchs the ctenidia are represented by muscular septa which pump water through the mantle cavity, and respiratory exchange occurs over part of the surface of the latter (figure 3·3) (p. 71).

Respiratory currents. In those archeogastropods which retain two ctenidia, water is normally drawn into the mantle cavity anteriorly on both sides of the head by the action of the lateral cilia on the gills. It passes over the gills and then out again through holes or a slit in the roof of the mantle cavity and the adjacent part of the shell. Thus in *Haliotis* the mantle cavity has a dorsal cleft, above which is a row of five holes in the shell, and the ctenidia divide the mantle cavity into ventral inhalant and dorsal exhalant chambers. In *Emarginula* both mantle cavity and shell have a dorsal cleft (figure 2·24), whereas in the keyhole limpet *Diodora* there is a single hole at the apex of the shell.

In *Pleurotomaria* the right gill is smaller than the left and the slit in the mantle cavity lies to the right of the head; in those archeogastropods which only have a left ctenidium (e.g. *Trochus*), and in the mesogastropods also, water enters the mantle cavity exclusively from the left side of the head and normally leaves on the right. The remainder of a slit in the right edge of the mantle persists in *Trochus* but not in any meso-gastropods. The anus and the nephridiophores open on the right side and so fouling of the inhalant current is prevented.

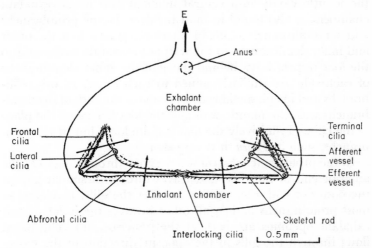

FIGURE 2·24 *Emarginula reticulata*. Section through the mantle cavity to show the arrangement and structure of the ctenidia. [After Yonge (1947) *Phil. Trans. Roy. Soc. B*, **232**]

The true limpets (Patellacea) are exceptional in that a pallial groove has been formed on either side of the foot by an over-hang of the mantle, i.e. a secondary mantle cavity. In the Acmaeidae there is a single, left ctenidium; in *Acmaea* itself this is the only gill, but in some other members of the family secondary gills are also present, lying in the pallial groove as in the chitons. This is carried a stage further in the Patellidae in that both ctenidia have been lost, and in *Patella* the secondary gills extend all the way round the pallial groove. In *Acmaea* water enters this groove to the left of the head and passes down both sides, leaving the animal at the hind end, but in those patellaceans which have developed secondary gills water enters the pallial groove between the gills.

In prosobranchs with pectinibranch gills the exhalant opening has also assumed an anterior position, water entering to the left and leaving to the right of the head. In a few mesogastropods and most neogastropods the mantle has become extended in the region of the inhalant aperture and its edges folded to form an inhalant siphon, which may be protected by a projection of the shell as in the whelk *Buccinum*.

The arrangement seen in *Haliotis*, in which the ctenidia divide the mantle cavity into ventral inhalant and dorsal exhalant chambers, is also found in most bivalves. In the protobranchs and some filibranchs, tufts of interlocking cilia join the outer and inner demibranchs to the wall of the mantle cavity and to the foot respectively. Behind the foot the inner demibranchs of each side are similarly joined to each other. In other filibranchs and in eulamellibranchs, fusion of the ends of the demibranchs with the mantle wall and the foot has taken the place of the cilia. (Primitively the foot is quite long, but in some filibranchs, such as *Pecten*, it is very short.)

In the protobranch *Nucula* water enters through a gap at the anterior end of the shell and leaves through a similar gap at the posterior end after passing up over the gills. However, in most protobranchs and in all other bivalves the inhalant and exhalant openings are both at the posterior end. The blood flows through the gills in the opposite direction to the movement of water over them, thus making use of the countercurrent principle discussed in chapter 1.

In protobranchs the respiratory current over the gills is produced by muscular movements of the filaments and, as in gastropods, by action of the lateral cilia. The gill movements reach their peak of development in the Nuculanidae (e.g. *Malletia*, figure 2·22*b*). In *Malletia* the muscular ctenidia are periodically raised and lowered, raising being accompanied by water entering the inhalant and leaving the exhalant opening. In filibranchs and eulamellibranchs there is no pumping action of the gills and the respiratory current is produced entirely by the lateral cilia. In eulamellibranchs water enters the vertical tubes formed as a result of the development of interlamellar and interfilamental junctions via the ostia.

In many species siphons have been developed—ventral, inhalant and dorsal, exhalant—and their development allows the animal to be almost completely buried in sand, mud or even rock. The siphons show considerable variation in different species: in some they are short, in others even longer than the body; they may be of equal or unequal length, and so on.

Small particles which pass through the gills in protobranchs are trapped in a secretion of the hypobranchial glands which lie above the exhalant chamber. In the other orders the gills are more tightly meshed and the laterofrontal cilia better developed, and these glands have disappeared.

In the septibranchs the septa formed from the ctenidia perform the dual role both of separating the ventral inhalant chambers from the dorsal exhalant chambers and of pumping water by muscular contractions from the former into the latter.

The respiratory currents of cephalopods, like septibranchs, are produced entirely by muscular contractions. In *Nautilus* water both enters and leaves the mantle cavity via a funnel (formed from the foot), and pulsations of this alone produce the inhalant and exhalant currents. In the Coleoida water still leaves the mantle cavity via the funnel, but enters between the funnel and the edges of the mantle. These currents are produced by contractions of the circular muscles of the mantle and the longitudinal muscles of the funnel and head. The exhalant current is also used for locomotion.

THREE
Lungs and Respiratory Trees

The efficiency of a respiratory system which utilises a small number of discrete sites, such as lungs, for the exchange of respiratory gases depends primarily on two factors. Firstly an increase in the efficiency of the uptake of oxygen itself—that is to say an increase in the degree and extent of vascularisation —and secondly the development of an efficient circulatory system which can rapidly transfer oxygen from its site of uptake to the relevant tissue, and can also remove the carbon dioxide and other waste products of cellular respiration to prevent their concentrations from building up in the tissues.

The definition of the term 'lung' is problematical. In a wide sense it may be used to include any specially adapted respiratory chamber whose walls are highly vascularised and form the primary region of oxygen uptake. This would include the mantle cavity of some molluscs and the branchial chambers of a number of crustaceans. However, the arachnid book-lungs and the respiratory trees of holothurian echinoderms (sea-cucumbers) would be excluded since these structures are in direct contact with the coelomic fluid rather than with fine branches of the vascular system. Possibly the presence of a tidal ventilatory mechanism (i.e. one in which expiration and inspiration are effected through a common opening), which these latter two groups of animals both possess, should be the main criterion; but this would then exclude the mantle cavity of prosobranch molluscs. Not only that, but it would include a number of other structures that would not normally be thought of as lungs. Thus in anisopteran dragonfly larvae, water is pumped in and out of a modified region of the hind-gut (the branchial chamber) through the anus; but the branchial chamber contains rows of tracheal gills which are the site of extraction of oxygen from the water (p. 50). In certain nemertines water is pumped in and out of the mouth, passing over a vascularised region of the fore-gut; but this area is not separated off in any way from the rest of the gut. There are also tidal mechanisms in two other groups of echinoderms, apart from the Holothuria. In some crinoids (feather stars) water is pumped in and out of the anus. In ophiuroids (brittle stars) the cilia of the ectodermal lining of the genital bursae cause respiratory currents, and water passes in and out of the slit-like genital

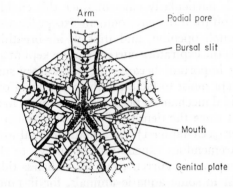

FIGURE 3·1 Oral view of the disc of *Ophiomusium* to show the five pairs of bursal slits. [After Hyman (1955) *The Invertebrates*, Vol. IV. McGraw-Hill: New York]

openings (bursal slits) of which, in most species, there is one pair at the base of each arm on the ventral (oral) surface (figure 3·1). However, in this last case the flow could be considered unidirectional, since the water passes into each bursa at the peripheral end of its opening and out at the oral end. Another unusual respiratory system, this time not tidal, is found in some phanerozonian starfishes. Here the nidamental chamber is utilised, water passing into it through marginal openings and spiracles and out through the osculum.

Thus as a collective term the word 'lung' is extremely difficult to define in a meaningful way. For the purposes of this book it will be taken to include the mantle cavity and branchial chambers of molluscs and crustaceans respectively, arachnid book-lungs and the respiratory trees of holothurians.

In arachnids, prosobranch and pulmonate molluscs and crustaceans the lungs are apparently a clear adaptation for a terrestrial existence, even though individual species possessing them may either have not completed the transition to a terrestrial habitat or have become secondarily aquatic. In a truly terrestrial animal gills would not function efficiently as respiratory structures since they tend to collapse and stick together in air and provide a large area over which water loss can occur (unless the animal can carry its own water supply about with it, as certain land crabs do, for example). A tidal ventilatory mechanism is particularly efficient since this enables the lung to be sealed off from the environment except for a single, comparatively small opening, and thus in air-breathing animals water loss via the expiratory current can be kept at a minimum. This is very important because the respiratory surfaces must always be kept moist to facilitate their uptake of oxygen. Fortunately a tidal mechanism becomes a possibility in a terrestrial environment since the density of air is very much lower than that of water (see chapter 1). An aquatic animal would have to undergo a tremendous expenditure of energy to rhythmically reverse its expiratory current; but nevertheless tidal mechanisms do exist in some aquatic animals, for instance in many echinoderms, nemertines and dragonfly larvae.

BOOK-LUNGS

3.1. Arachnida Book-lungs are an exclusive feature of certain members of the class Arachnida (phylum Arthropoda). This is a particularly interesting group of animals in that in some orders the respiratory organs are book-lungs, in others tracheal systems, while in the Araneida (true spiders) many species possess both of these respiratory structures. This, together with the fact that in most spiders that have tracheae the openings occur on the same segment or segments as do those of the book-lungs in other species, has led to the supposition that these structures are homologous and that tracheae, in this order at least, have evolved from book-lungs.

In their turn book-lungs may well have been evolved from book-gills, possibly similar in nature to those found in living merostomes (king-crabs) (chapter 2). This could have occurred by invagination of the book-gills to form internal structures at the time the arachnids invaded the land.

The body of arachnids is divided into two distinct zones, an anterior prosoma and a posterior opisthosoma. Book-lungs are essentially similar in the members of all six of the orders that possess them—Scorpionida, Palpigradida, Uropygi, Amblypygi, Schizomida and Araneida. They are paired and the openings (spiracles) leading into them always open onto the

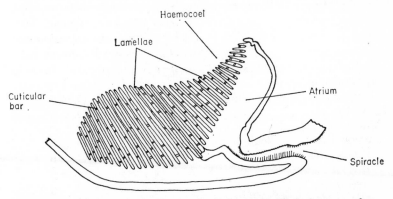

Figure 3·2 Section through the book-lung of *Thelyphonus caudatus* (Uropyge). [After Millot (1949) in *Traité de Zoologie*, Vol. VI, (Ed. Grassé). Masson: Paris (after Börner)]

opisthosoma on one or more of the second to the sixth segments. (In merostomes (chapter 2) there is a pair of book-gills on each of the second to the sixth opisthosomal segments.) Each spiracle opens into an atrium, from one of the inner walls of which arises a series of flat plates or lamellae. Adjacent lamellae are held apart by cuticular bars (figure 3·2). The lumina of the lamellae are continuous with that of the atrium and so air can circulate through them. The number of lamellae varies: it can be as few as four or five (e.g. *Dictyra* (Araneida)) or as many as 150 (e.g. *Liphistius* (Araneida)). They lie in the haemocoel and are bathed by the circulatory fluid, and as a result diffusion of respiratory gases occurs across their walls.

Respiratory efficiency has been improved in two ways—by control of the spiracles, and by the development of muscular ventilating movements of parts of the body. Some scorpions and spiders for example possess muscles that can expand the spiracles and atrium, but this mechanism is only brought into play during periods of great activity. Spiracular control is presumably of at least equal importance in the conservation of water.

In the Scorpionida (scorpions) there are four pairs of book-lungs, opening on the third to the sixth opisthosomal segments

TABLE 3.1. Respiratory openings in mesostomes and arachnids

	Respiratory organ	No. of pairs (when present) SM=Single median	Segments (O=Opisthosoma P=Prosoma)
Merostomata			
Eurypterida	Book-gills	6	O. 1–6
Xiphosura	,,	5	O. 2–6
Arachnida			
Scorpionida	Book-lungs	4	O. 3–6*
Palpigradida	,,	absent or 3	O. 4–6
Uropygi	,,	2	O. 2–3
Amblypygi	,,	2	O. 2–3
Schizomida	,,	1	O. 2
Araneida	Book-lungs	1 or 2	O. 2 or 2–3
	Book-lungs + Tracheae	2 or 1 + SM	O. 2–3 or 2+various
	Tracheae	2	O. 2–3
Pseudoscorpionida	Tracheae	2	O. 3–4
Phalangida	,,	2	O. 4
Ricinuleida	,,	1	P. 6
Solifugae	,,	3 or 3 + SM	O. 3–5 or O. 3–5+P4
Acarina	,,	†	†

* Really 4–7 because first opisthosomal segment is missing
† Very variable; see table 4.1

(really segments four to seven, because the first opisthosomal segment is missing in scorpions) (table 3.1). They are absent in the majority of the Palpigradida, but when they do occur there are always three pairs, one on each of the fourth to the sixth opisthosomal segments. In the Uropygi, Amblypygi and Schizomida (collectively known as whip-scorpions) there is a pair on the second segment of the opisthosoma, and in the first two of these orders there is an additional pair on the third segment. Also in a few spiders (Araneida), such as *Atypus*, there is a pair on each of the second and third opisthosomal segments, but in most spiders those on the third segment have been replaced by tracheae. In some cases the tracheal spiracles have fused in the ventral mid-line, and this median spiracle may remain on the third segment or occupy a more posterior position. In certain spiders there is only a single pair of spiracles, in which case it is always the anterior pair, and these invariably open into book-lungs.

LUNGS

Lungs have evolved on a number of occasions in both molluscs and crustaceans. In all but one instance (the septibranch molluscs) their development coincides with a more or less successful invasion of the terrestrial environment.

3.2. Crustacea Lungs occur in two orders of the subclass Malacostraca, the Isopoda and the Decapoda.

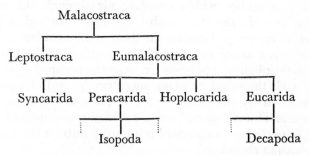

The Isopoda includes the woodlice and in one family of these animals, the Oniscoidea, both amphibious and terrestrial species can be found. In this family, the exopodite of each abdominal appendage forms an operculum and in the truly terrestrial species of the Oniscidae the opercula contain lung-like cavities.

Among the decapods, amphibious and terrestrial species occur in only two of the pleocyematan infraorders—the Ano-mura (hermit crabs) and the Brachyura (true crabs); there are no terrestrial dendrobranchiatans.

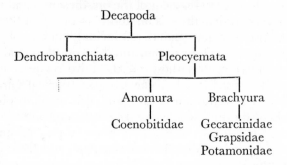

Four families are involved, the Coenobitidae in the Anomura, and the Gecarcinidae, Grapsidae (some) and Potamonidae (some) in the Brachyura. All are tropical, retain gills (although these are generally considerably reduced in number and sur-face area) and return to the water to breed. The potamonids differ from the others in that they breed in fresh water.

The initial stages in the invasion of the land can be seen in amphibious species, which still retain some water in their branchial chambers. This may be aerated by movements of the scaphognathites which circulate air through the chambers (e.g. *Ocypode*, the ghost crabs). Air enters through a posterior pair of openings (between the third and fourth legs in *Ocypode*) and leaves anteriorly through the openings at the bases of the claws (inhalant apertures in aquatic decapods) as well as through the usual exhalant apertures (p. 45). Respiration is improved by the development of vascularised areas in some species. Thus *O. ceratophthalma* has highly vascularised gills and *O. quadrata* has vascularised respiratory tufts on the walls of the branchial chambers.

Terrestrial species, unlike the amphibious ones, do not carry water in their branchial chambers. The branchial chambers tend to become enlarged, and the number of gills and their surface area reduced still further. Some gecarcinids have highly vascularised gills, and in both gecarcinids and grapsids there is a vascular network on the median wall of each branchial chamber. The most elaborate respiratory modifications are seen in coenobitids: in *Coenobita* the floor of each chamber is vascularised, while in *Birgus* (the coconut or robber crab) each branchial cavity is divided into an upper and a lower region by a septum. The gills are in the lower part and the inner walls and roof of the upper chamber are protruded into a series of vascularised feathery tufts. Ventilation takes place in *Birgus* by movements of the scaphognathites and by raising and lowering the pleural and posterior margins of the carapace. Raising the posterior margin of the carapace also occurs in gecarcinids and grapsids.

3.3. Mollusca In protobranch, filibranch and eulamellibranch bivalves there is a single gill hanging from the roof of the mantle cavity on each side of the body (figures 2·22–2·24). However, in septibranchs these gills are replaced by muscular septa which pump water through the mantle cavity (figure 3·3).

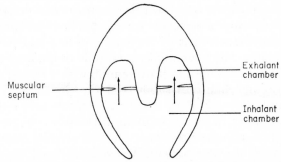

Muscular septum

Exhalant chamber

Inhalant chamber

FIGURE 3·3 Section through a septibranch bivalve to show the modification of the 'gills' into muscular septa. Arrows indicate direction of respiratory current

Respiratory exchange occurs at the surface of the mantle cavity above the septa (epibranchial or exhalant chamber) and so the mantle cavity serves as a primitive sort of lung, although this is not associated with any apparent attempt to invade the terrestrial environment.

In gastropods a lung has evolved on at least two separate occasions, i.e. in the prosobranchs and in the pulmonates, and is associated in each case with invasion of the terrestrial environment. In both groups the mantle cavity has become modified to form a lung, the ctenidia have been lost, and inspiration and expiration occur through a common aperture (tidal ventilation).

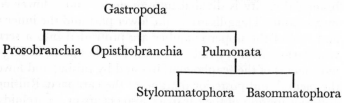

Among the prosobranchs there are a number of amphibious species and these retain at least one ctenidium. In the Ampullariidae (*Ampullarius* and *Pila*) the mantle cavity is partially divided by a ridge on its floor into right and left halves, a single ctenidium lying on the right and an opening on the left side leading into a vascularised lung. These animals live in stagnant water and breath air. This is drawn into the mantle cavity through an inhalant siphon which penetrates the surface of the water. Ventilatory movements are effected by rhythmic inward and outward movements of the head and foot and pulsations of the mantle. The animal may retain air in the lung when it leaves the surface and thus provide itself with a second method of respiration while it is submerged (i.e. in addition to the ctenidium). Not only that, but the air bubble may serve as a buoyancy tank. In the shore-living littorinids a single monopectinate ctenidium is retained, but vascularised folds of the mantle wall are also present. When submerged the mantle cavity is filled with water; when exposed it is filled with air.

Invasion of the land has occurred on more than one occasion in the prosobranchs, and terrestrial species are found in the archeogastropods (Helecinidae) and the mesogastropods (Cyclophoridae and Pomatiasidae). These molluscs are all operculate, have lost both ctenidia and their air-filled mantle cavity has a vascularised roof. In some tropical species a small secondary opening into the mantle cavity exists (figure 3·4) and this is used for ventilation when the operculum is withdrawn, such as during periods of draught. This enables respiration to continue without serious water loss.

Most pulmonates are primarily terrestrial, or have passed through a terrestrial stage in their evolution. In the majority of those species which are still truly terrestrial the mantle cavity is anterior in position and has only a small contractile opening (the pneumostome) to the exterior. The ctenidia have been

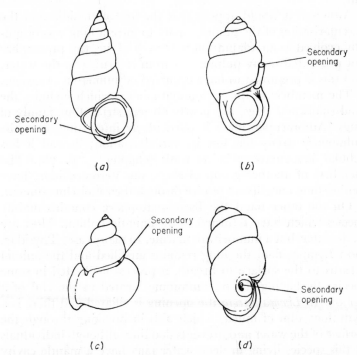

(a) (b)

(c) (d)

FIGURE 3·4 Secondary openings into the mantle cavity of some tropical terrestrial prosobranchs. (Cyclophoridae) (a) *Tortulosa pyrimidatus;* (b) *Rhaphaulus chrysalis;* (c) *Alycaeus major;* (d) *Pupinella macgregori.* [After Purchon (1968) *The Biology of the Mollusca.* Pergamon Press: Oxford]

lost and the mantle wall is covered with ridges which contain pulmonary veins, the net effect being to produce a highly vascularised surface. The muscular floor and walls can be moved rhythmically to pump air in and out of the lung via the pneumostome, which opens and closes synchronously. During inspiration the pneumostome is opened and the floor of the mantle cavity lowered. The pneumostome is closed and the floor muscles relaxed to allow the floor to return to its raised position. This leads to an increase in pressure within the lung and consequently aids in the uptake of oxygen. Expiration is accomplished simply by opening the pneumostome, whereupon the increased pressure inside causes the air to be forced out.

Although it would appear that the loss of ctenidia and the vascularisation of the mantle cavity to form a lung was originally evolved as an adaptation to terrestrial life, this process has not prevented many pulmonates from returning to the water, and this is presumed to have occurred on numerous occasions.

The members of the Stylommatophora, which includes the land snails and slugs, are nearly all terrestrial. One family of slugs (Athoracophoridae) is particularly interesting since the pulmonary cavity has lost its vascularisation; instead it has tubular invaginations in its walls (chapter 3, p. 78). The members of another group of slugs, the Veronicellidae, have neither lung nor gill and respire through their moist integument.

On the other hand the Basommatophora contains mainly species which have returned to an aquatic habitat. Most are fresh-water, but a number are marine. In many cases (Physidae, most Lymnaeidae) the lung remains air-filled and the animal returns to the surface to breath, a process facilitated in some species by having the pneumostome located at the end of a siphon. In *Limnea peregra* the opening is ciliated and there is a hydrofuge film of mucus which aids in breaking through the surface of the water and prevents flooding, although individuals of this species living in deep water may have a mantle cavity filled with water.

Many species actually pump water in and out of the lung, but these tend to possess secondary gills either inside or outside the mantle cavity. Thus in the siphonariid limpets and the Amphibolidae (both marine families) the mantle cavity contains secondary or substitute gills. *Siphonaria* for example has a series of up to 30 gill leaflets which hang down from the roof of the mantle cavity. In others, such as the Ancylidae (fresh-water), there is a reduced mantle cavity and a single secondary or adaptive gill (pseudobranch) is present in the form of a conical projection from the left side of the foot, i.e. exterior to the mantle cavity.

It is somewhat difficult to envisage a terrestrial ancestry for marine limpets such as *Siphonaria*, and indeed it is thought that they represent a very early offshoot from the main gastropod stock which never became entirely terrestrial.

RESPIRATORY TREES

3.4. Echinodermata Respiratory trees are found exclusively in one class of echinoderms, the Holothuroidea, and within this class they occur in three of the orders. There are two respiratory trees, one on the right the other on the left. They are long,

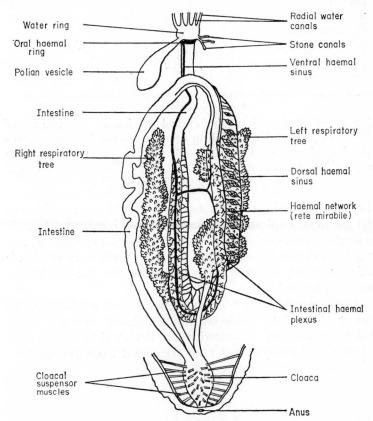

FIGURE 3·5 Respiratory trees and the circulatory system of *Holothuria tubulosa*. [After Cuenot (1948) In *Traité de Zoologie*, Vol. XI, (Ed. Grassé). Masson: Paris]

branched tubular structures which open into the cloaca (figure 3·5), and pumping movements of the latter cause water to flow in and out of the trees in a tidal fashion. The branches end in thin-walled ampullae which lie in the coelomic fluid, and it is presumably through these that the bulk of gaseous exchange occurs.

Inspiration is effected by contraction of the dilator muscles of the cloaca, causing water to flow into it. The anus is then closed, the cloaca contracts, and the valve between cloaca and respiratory trees opens; the water is thus forced into the trees.

FIGURE 3·6 A record of the cloacal respiratory movements of *Holothuria forskali*. Each small upward movement indicates a contraction of the cloaca; the large downward movement shows the rapid ejection of water. [From Newell (1970) *The Biology of Intertidal Animals*. Logos: London]

In *Holothuria* this process is repeated from six to ten times to fill the trees. Expiration on the other hand is achieved by only a single contraction of the trees, when water is expelled through the anus via the cloaca (figure 3·6). The ventilatory rhythm is fairly regular, the frequency depending on the respiratory needs of the animal.

In some genera the lower branches are not respiratory, but are modified into Cuvierian tubules. These secrete a sticky substance and can be extruded and detached to form a mass of elongated sticky threads; this acts as a defence mechanism.

FOUR
Tracheal Systems

Tracheae are to be found only in members of two invertebrate phyla, the Arthropoda and the Onychophora, although similar respiratory tubules do occur in a few molluscs and crustaceans. They are primarily an adaptation to the terrestrial environment, but a number of species possessing them have become secondarily aquatic. Among the arthropods they are typically found in most insects, chilopods (centipedes), diplopods (millipedes) and symphylans, as well as in a number of arachnid orders.

Tracheae are invaginations of the body surface, normally communicating with the exterior via a hole or spiracle. In some aquatic insect larvae there are no spiracles and the system is said to be 'closed'. Except in primitive forms the spiracle opens into an atrium, and the tracheae lead off from this. The tracheae may be short and unbranched and end in the haemocoel, and thereafter reliance is placed on the circulatory system to transport the oxygen to the tissues. Such a structure is called a tracheal lung (figure 4·1) and is similar in some ways to the book-lung described in the previous chapter (p. 66) although it is tubular rather than lamellate. Tracheal lungs are found in some arachnids and in scutigeromorph centipedes.

In most cases the tracheae are long and terminate in small, fluid-filled tubules (tracheoles) which are in intimate contact with the tissues of the body. There are two basic types—sieve tracheae and tube tracheae. In the former a great number of tracheae arise from the atrium, whereas in the latter there is only a single unbranched or branched tube. The tracheae are lined with chitin and are extracellular structures but the terminal tracheoles however are intracellular.

There are several ways in which a tracheal system can become more efficient. In many cases quite elaborate spiracular closing mechanisms have developed which allow a greater degree of independence from the environment, in that water losses via the spiracles can be better controlled. Branching of the tracheae increases the capacity of the system for holding air and the amount of oxygen reaching the tissues. Connections between the tracheae on each side of the body (commissural linkage) and between adjacent segments (longitudinal linkage) mean that damage to any one spiracle does not lead to oxygen deprivation of any tissue. Furthermore longitudinal linkage, together with the development of a ventilating mechanism and the sequential opening and closing of specific spiracles, allows the possibility of an unidirectional airflow through the system (chapter 8). In animals which actively ventilate their tracheal system, efficiency can be further increased by the development of air sacs, which are essentially dilations of the tracheae.

Apart from the true tracheae, respiratory tubules are found in two other groups. In certain terrestrial isopods (Crustacea) belonging to the oniscoid families Porcellionidae and Armadillididae, tube-like invaginations occur in either the first or all five opercula (modified exopodites of the abdominal appendages) (chapter 2). These are commonly referred to as pseudotracheae. In one family of slugs (Athoracophoridae) the vascularisation of the pulmonary mantle cavity has been lost and invaginations of the walls form trachea-like tubules which penetrate the surrounding tissues and extend into the blood sinuses.

TRACHEAL LUNGS

4.1. Chilopoda In the Chilopoda (centipedes) tracheal lungs are present in only one order—the Scutigeromorpha (cf. p. 81). The tergal plates over the fifteen leg-bearing segments of these animals are large and reduced in number to seven. There is a longitudinal, slit-shaped spiracle on each one, situated mid-dorsally towards the posterior edge of the plate. Each spiracle opens into an atrium from which arises a large number of short tracheal tubes. These are arranged fan-like on either side of the mid-line and are immersed in the blood of the pericardial cavity (figure 4·1a).

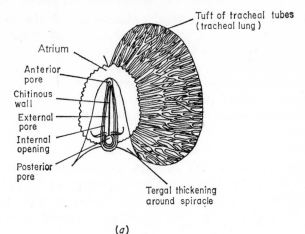

(a)

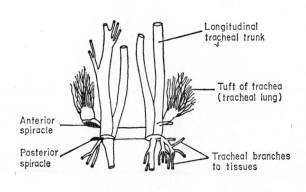

(b)

FIGURE 4·1 Tracheal lungs of (a) a scutigerid centipede and (b) a spider (Arachnida, Araneida). [After Meglitsch (1967) *Invertebrate Zoology*. Oxford University Press: Oxford]

4.2. Arachnida It is only among the true spiders (Araneida) that tracheal lungs occur. In caponiid spiders for instance (e.g. *Nops coccineus*) there are two pairs of spiracles, one pair on each of the second and third opisthosomal segments. Each spiracle opens into an atrium which has a relatively thick lining of cuticle. The posterior pair are invariably associated with tube tracheae, but the anterior pair open into tracheal lungs; in this case a tuft of short tracheae arises from the atrium and lies in a sinus of the haemocoel. Tracheal lungs also occur in a few other species, generally associated with the posterior pair of spiracles, as for example in those spiders in which the anterior pair open into book-lungs (figure 4·1*b*).

4.3. Insecta Tracheal lungs are apparently present in a few species. They are said to occur for example in the larvae of *Hypoderma* (Diptera) where the tracheae divide at their ends into numerous branches which terminate in the body cavity.

TRACHEAE

4.4. Onychophora The spiracles are numerous all over the body. The openings are very small and each leads into a short, tube-like atrium lined with a delicate cuticle from the base of which a tuft of tracheae arises (figure 4·2). Normally each one is

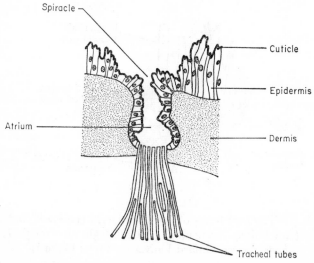

FIGURE 4·2 Spiracle, atrium and tracheal tubes (unbranched) of an onychophoran. [After Pflugfelder (1955) *Mikrokosmos*, **44**]

a simple, unbranched tube which extends direct to the tissue which it supplies, but in some species the walls of the tracheae have spiral chitinous supports. Although the body wall is more impermeable to water than say that of an earthworm, the water loss is about twice as great. This is presumably due to the lack of any spiracular control, which results in the spiracles remaining open at all times. Inevitably onychophorans, which incidentally are all terrestrial, have to remain in moist surroundings to avoid rapid dessication.

4.5. Diplopoda The Diplopoda (millipedes) is one of four classes often grouped together as the Myriapoda. The other three are the Symphyla, the Pauropoda (which are all very small and do not possess specialised respiratory structures) and the Chilopoda (centipedes). The members of these classes are all mandibulates but, unlike insects, the body is only divided into two tagmata, the head and the trunk.

The millipedes are characterised by the fusion of adjacent pairs of trunk somites to form double segments or diplosomites, each bearing two pairs of legs. On all but the first three trunk segments there are four spiracles (two pairs), one just anterolateral to the coxa of each of the four legs. The spiracles lead into air reservoirs which are situated in the transverse apodemes to which the leg muscles attach and from each reservoir a tuft of tracheae arises. The tracheae are normally unbranched and extend directly to the tissues as in onychophorans. The tissues of the first three trunk segments receive tracheae from the reservoirs of the fourth segment.

4.6. Symphyla Unlike millipedes there is no fusion of the trunk segments to form diplosomites in symphylans. There are only 12 leg-bearing trunk segments, each with one pair of legs, and superficially they look rather like centipedes. They only possess a single pair of spiracles, one located on either side of the head, and the tracheae are branched and supply both the head and the first three trunk segments.

4.7. Chilopoda The tracheal system in centipedes is more highly developed than in any of the other myriapod groups. All, with the exception of the scutigeromorphs (p. 78), possess tube tracheae. The spiracles, of which there is primitively one pair per somite, lie laterally in the pleural region, above and slightly behind the coxae of the legs. In geophilids the spiracles are only absent on the first and last somites, but the pattern varies considerably in other groups and in many cases a number of somites are devoid of spiracles. Lithobiomorphs for example usually have no spiracles on the somites with narrow tergal plates.

The geophilids are also somewhat primitive in that the spiracles are relatively simple and longitudinal tracheal trunks are absent. In other chilopods branching and anastomosis of tracheae are common and longitudinal trunks present.

It is of interest to note that, contrary to the generally accepted notion that all myriapods are terrestrial, a few species of geophilids are in fact intertidal, and are submerged at high tide. On the whole they probably retain sufficient air within the tracheae to last during this period. However, in the Indian species *Myxophilus indicus*, and possibly in some others, additional air is both trapped on the surface of the coxae and lodged as a bubble in the curved end of the trunk.

4.8. Arachnida As we saw in the previous chapter some orders of arachnids possess book-lungs and others have tracheal systems, while the true spiders (Araneida) have either or a combination of both.

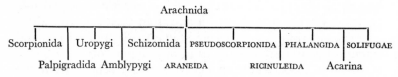

The following, with the exception of some small species and some araneids, possess tracheae: Pseudoscorpionida, Ricinuleida, Phalangida, Acarina, Araneida and Solifugae. Whereas book-lungs always occur on the opisthosoma—the arachnid body is divided into an anterior prosoma and a posterior opisthosoma—in some instances tracheae open onto the prosoma. Each spiracle opens into a vestibule (atrium) which is lined with a thick layer of cuticle, as in the case of tracheal lungs, and this is continued in the tracheae, since they originate from invaginations of the ectoderm. The pseudoscorpions, ricinuleids and some spiders possess sieve tracheae; they always possess an atrial chamber and from this arises a large bundle of tracheae.

Pseudoscorpions (false scorpions) have two pairs of spiracles, located on the third and fourth opisthosomal segments. The tracheae from the anterior pair run forwards; those from the posterior pair backwards. In the Ricinuleida there is only a single pair of spiracles and these are located on the dorsolateral angles of the posterior margin of the prosoma, just above the coxae of the third legs. Each atrium gives rise to hundreds of very small unbranched tubes that directly supply the various prosomal organs.

In the other three groups and in most spiders the tracheae are simple unbranched or branched tubes (tube tracheae) usually, but not always, originating from an atrium. In the more highly evolved systems both commissural and longitudinal linkage occur. The phalangids (harvestmen) have one pair of spiracles which lie on either side of the second opisthosomal sternite at the base of the last pair of legs. Each gives rise to a large, longitudinal trunk which passes forwards into the prosoma, sending branches to the legs and prosomal organs. There are also branches to the opisthosoma, but these are not so well developed. In the very active, long-legged harvestmen secondary spiracles occur on the tibia of the legs.

The Acarina can be broadly divided into mites and ticks: the former possess simple tracheae, but in the latter extensive branching to all parts of the body occurs. When present there are between one and four pairs of spiracles, the position of which varies considerably. In only one group, the Tetrapodili, is a tracheal system always absent.

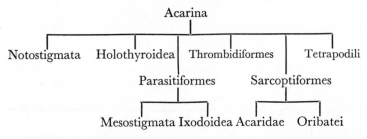

In the Notostigmata there are four pairs of spiracles, dorsally situated on the opisthosoma, and this is the only group to bear spiracles on this tagma of the body. In the Holothyroidea there are two pairs, one situated on the border of a dorsal shield beneath the coxae of the third pair of legs, the other (which lead into air sacs) lying on the margin of the same shield but slightly behind the fourth pair of legs.

Members of the Parasitiformes possess only a single pair of spiracles, usually laterally or lateroventrally. There are two sections in this suborder: the Mesostigmata and the Ixodoidea. In the former the spiracles lie at the extreme posterior ends of a pair of projections called the peritremes, and are normally found just behind the coxae of the third legs (the peritremes

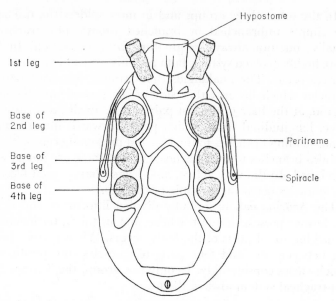

FIGURE 4·3 Ventral view of a female mesostigmatan to show the peritremes. [After André (1949) In *Traite de Zoologie*, Vol. VI, (Ed. Grassé). Masson: Paris]

extending from about the level of the second legs (figure 4·3)). However, the peritremes are sometimes reduced and occasionally absent, as for example in the Ixodoidea (ticks) in which there are never any peritremes. There are two main superfamilies of ticks: the Ixodei (hard ticks), in which the spiracles are situated lateroventrally behind the coxae of the fourth legs and consist of a porous spiracular plate; and the Argasides (soft ticks) in which there is no such plate and the spiracles lie laterally near the coxae of the third or fourth legs. The tracheae have a spiral thickening of chitin.

The Thrombidiformes presents us with an even more complex picture. In one group for example (Heterostigmata) there is sexual dimorphism, the tracheal system being absent, or at best rudimentary, in the males, whereas the females have a functional tracheal system with a single pair of spiracles opening laterally behind the pedipalps. In some Prostigmata spiracles are absent, in others they are present. The few that have been studied all possess a single pair, with one spiracle on each side of the dorsal mid-line, near the chelicerae. The two principal tracheal trunks usually each give rise to a tuft of fine, ramifying tracheal branches (figure 4·4) and in a few species the branches have a spiral thickening.

The Sarcoptiformes are divided into the Acaridae and the Oribatei. In the former spiracles are absent; in the latter they are either absent (Ptychina) or there are four pairs (Aptychina).

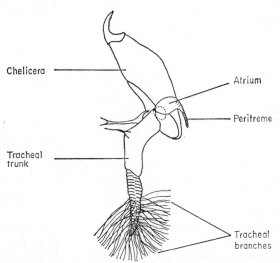

Chelicera

Atrium

Peritreme

Tracheal trunk

Tracheal branches

FIGURE 4·4 Chelicera of *Allothrombium fuliginosum* (Thrombidiformes —Prostigmata) with peritreme at its base. [After André (1949) In *Traité de Zoologie*, Vol. VI, (Ed. Grassé). Masson: Paris]

In the Aptychina the coxae form a coxal plate and there is a spiracular opening between the edge of this plate and the trochanter of each leg.

The disposition of the spiracles in the various suborders of acarines is summarised in table 4.1.

In the true spiders (Araneida) the respiratory apparatus may be in the form of book-lungs, tracheae, or both on the same individual. Nearly all possess two pairs of spiracles which open on the ventral surface of the opisthosoma, one on the second and the other normally on the third segment. In the majority (e.g. Argiopoidae) the first pair open into book-lungs, the second pair into tracheae. However, in some there are two pairs of book-lungs (e.g. Atypidae), and in others two pairs of

TABLE 4.1. Spiracular arrangements in the acarines

	No. of pairs	Opisthoma (O) or Prosoma (P)	Position
Notostigmata	4	O	Dorsal
Holothyroidea	2	P	Near coxae of legs 3 and 4
Parasitiformes			
Mesostigmata	1	P	Lateral or latero-ventral. At end of peritreme near coxae of legs 3 or 4
Ixodoidea			
Ixodei	1	P	Latero-ventral. On spiracular plate behind coxae of legs 4
Argasides	1	P	Lateral. Near coxae of legs 3
Thrombidiformes			
Tarsonemini	Absent or 1*	P	Lateral. Behind pedipalps
Prostigmata	Absent or 1	P	Dorsal. Near base of chelicerae
Sarcoptiformes			
Acaridae	Absent	—	—
Oribatei			
Ptychina	Absent	—	—
Aptychina	4	P	Between coxal plate and trochanter of each leg
Tetrapodili	Absent	—	—

* Only in females

tracheal openings (e.g. Caponiidae). In a few families, such as the Lycosidae, the hind pair of spiracles have fused in the midline and the resulting single spiracle has become sited posteriorly towards the hind end of the opisthosoma. In all such cases this single median spiracle opens into a tracheal system, while

PLATE I

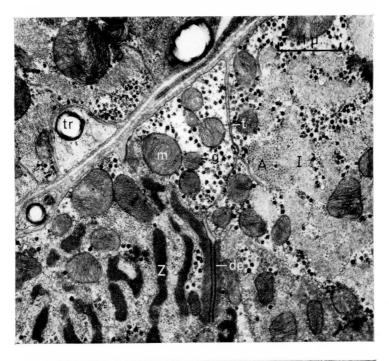

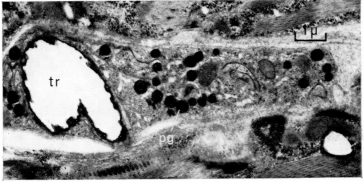

FIGURE 4·6 Electronmicrographs of the expiratory dorsoventral muscle of an aeshnid dragonfly larva showing tracheoles (*tr*) *A*, A-band, *de*, desmosome; *I*, I-band; *m*, mitochondrion; *p.g.*, pigment granules; *t*, T-system tubule; *Z*, Z-line. [From Mill and Lowe (1971) *J. ins. Physiol.* **17**]

PLATE II

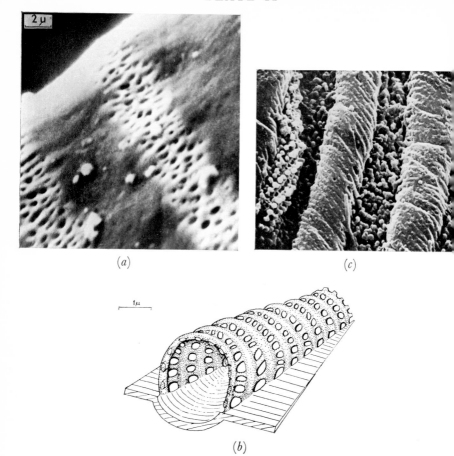

(a)

(c)

(b)

FIGURE 4·12 (a) Stereoscan electronmicrograph of the flat plastron lines of the gill of *Lipsothrix remota*. (b) Diagram and (c) stereoscan electronmicrograph of the arched plastron lines on the gill of *Antocha bifida*. [a from Hinton (1967) *Proc. R. ent. Soc. A*, **42.** b, c from Hinton (1966) *Proc. R. ent. Soc. A*, **41**]

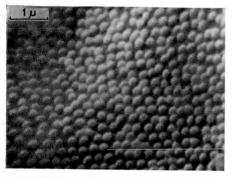

FIGURE 4·13 Stereoscan electronmicrograph of the plastron on the gill of *Orimargula hintoni*. [From Hinton (1968) *Advances in Insect Physiology*, (Ed. Beament, Treherne and Wigglesworth). Academic Press: London]

PLATE III

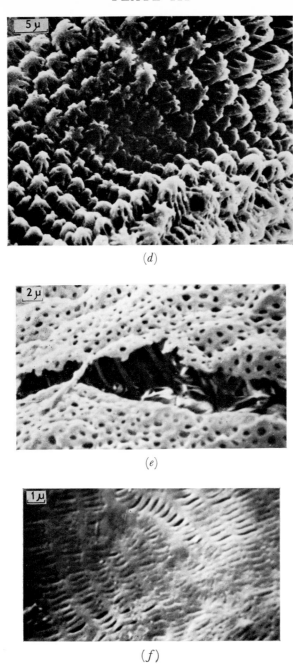

(d)

(e)

(f)

FIGURE 4·15 *continued.* Electron micrographs of the plastron of (d) *Geranomyia unicolor,* (e) *Dicranomyia monostromia,* (f) *Aphrosylus celtiber.* [All from Hinton (1968) *Advances in Insect Physiology,* **5**]

the anterior pair always opens into book-lungs, as they also do when the hind pair have completely disappeared (e.g. Pholcidae). There is little or no correlation between the pattern of respiratory organs and the classification of the araneids.

Where a tracheal system is present in the true spiders each spiracle normally opens into an atrium. The maximum degree of development is reached in such families as the Dysderidae, in which dense tufts of tracheae invade all the organs of the body. In others (e.g. Dictynidae) the system is rather less extensive, but there are still numerous tracheae and these penetrate into the prosoma and its appendages. However, in many families the tracheae remain simple along their length (with very few branches) and are restricted to the opisthosoma. In the Pholcidae the spiracles have disappeared and so the tracheal system is no longer in communication with the exterior.

In a few species of araneids the interior surface of the tracheae is similar to that found in insects in that they possess a spiral thickening. However, more often they are lined with numerous small spines which may be interwoven to form a network.

The Solifugae (sun scorpions) have a very highly developed tracheal system, presumably associated with their very active life and capacity for very rapid movements. They have three pairs of ventral spiracles: one pair is on the prosoma, just posterior to the coxae of the second pair of legs; the other two pairs open onto the opisthosoma between the third and fourth and between the fourth and fifth segments respectively. In some cases there is an additional single, median spiracle lying behind the fifth sternite. The main tracheal trunks are very large and give off branches which ramify to all parts of the body.

4.9. Insecta The vast majority of insects possess a tracheal system. However, a few smaller ones (e.g. most Collembola), some early aquatic stages and the larvae of some endoparasites rely exclusively on surface respiratory exchange. With the possible exception of the embryo of *Apis* and a few collembolans, spiracles are restricted to the thorax and abdomen. Embryologically there are thought to be twelve pairs, three on the thorax and nine on the abdomen, but in most embryos the prothoracic pair is absent, and often the ninth abdominal pair

D

also. They are normally lateral in position. The respiratory system in insects can be subdivided on the basis of the number and arrangement of the spiracles:

(a) Holopneustic: 10 functional pairs of spiracles (2 thoracic and 8 abdominal).
(b) Hemipneustic: less than 10 functional pairs (can be further subdivided on the basis of which pair or pairs are non-functional).
(c) Apneustic: no functional spiracles.
(d) Hypopneustic: one or more pairs disappeared (as distinct from become non-functional).

Typically each spiracle is surrounded by an area called the peritreme and opens into an atrium (figure 4·5), although the latter is sometimes absent (e.g. *Sminthurus*). The tracheae are tubular invaginations of the atrium and since they are extra-

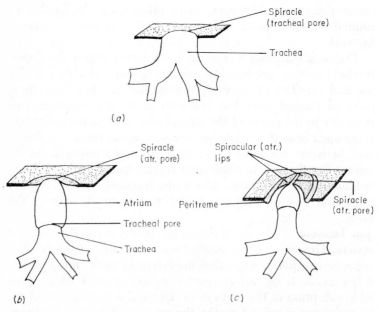

FIGURE 4·5 Diagrams showing (*a*) a simple spiracle, (*b*) the presence of an atrium and (*c*) the development of movable lips and a peritreme. *atr.*, atrial. [After Meglitsch (1967) *Invertebrate Zoology*. Oxford University Press: Oxford]

cellular structures they are lined with cuticle which is shed at ecdysis. This is normally a comparatively simple process in which the remains of the old cuticle and tracheal lining are pulled out through the newly formed spiracle. However, in some cases the aperture of the spiracle, or the atrium, contains structures which prevent this, and a temporary ecdysial opening or ecdysial tube is formed. The cuticle is thickened internally to form ridges (taenidia), which may exist as separate rings around the trachea, but more often form spirals which are periodically interrupted. Their function is to keep the tracheae distended and so allow the free passage of air. Surrounding the cuticular layer there is a pavement epithelium and then a thin basement membrane.

The tracheae terminate in intracellular structures called tracheoles which ramify between the cells (or in some instances, such as the flight muscles of certain insects, within the cells) of the tissues. There is no cuticular lining, but ridges are still present and can be seen under the electron microscope (figure 4·6; plate I).

The narrower parts of the tracheae and the intracellular tracheoles are filled with fluid. During cellular respiration metabolites tend to build up in the tissues and increase the osmotic pressure of the tissue cells, and presumably also that of the surrounding blood and tracheolar cells. This causes the fluid in the tracheal system to be withdrawn by the tracheolar cells, and thus bring the air/water interface into the close proximity of the active tissue (figure 4·7). This in turn reduces the diffusion time of the oxygen to the tissues. When activity decreases and the excess metabolites can be oxidised and removed by the vascular system, the osmotic pressure is lowered and more fluid enters the tracheoles again. This is an example of a self-regulating or homeostatic system.

The tracheal system in insects has evolved along three main lines: fusion of the tracheal trunks; provision of filters to keep out dust and restrict water loss; and the development of spiracular or atrial closing mechanisms. In many apterygotes the tracheae arising from each spiracle are unconnected with those from any other spiracle, but in most insects a series of transverse and longitudinal trunks has developed. There are three main divisions of each trachea—dorsal, visceral and ventral—and the dorsal and ventral ones may unite with those of the opposite side to form transverse trunks (commissures). The most common longitudinal trunks are the lateral ones but dorsal ones are often present, connected to the lateral ones with palisade tracheae. Occasionally there are ventral trunks also.

In a number of insects the tracheae are expanded in certain places, with concurrent loss of the taenidia, to form thin-walled

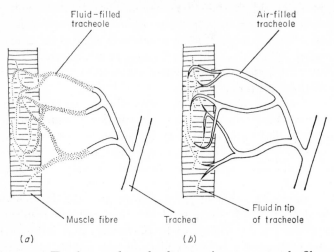

FIGURE 4·7 Trachea and tracheoles running to a muscle fibre, (*a*) with the muscle at rest and (*b*) with the muscle fatigued. [After Wigglesworth (1953) *The Principles of Insect Physiology*. Methuen: London]

air sacs. These occur for example in the larva of *Eristalis* and in many flying insects, reaching their greatest degree of development in the Diptera and Hymenoptera. Primarily they serve both to increase the volume of tidal air which can be changed during respiration and to decrease the diffusion path to the tissues (figure 4·8); also in animals in which they are found there is normally a unidirectional flow of air through part of the tracheal system. In addition they may serve as air stores, as in the littoral carabid beetles *Aepus* and *Aepopsis*. In some cases they even have non-respiratory functions, such as decreasing the specific gravity of flying insects and thus making the flight of large species easier; acting as hydrostatic organs in aquatic larvae; and increasing the efficiency of the tympanal (hearing) organs by allowing the tympanal membrane to vibrate freely.

The energy requirements for insect flight muscle are the highest known for any active tissue; values between 1.4 and 7.3 ml O_2/g muscle/m having been obtained for actively flying insects. This means that a considerable exchange of air between the atmosphere and the wing muscles must occur. Since the tracheal supply to the flight muscles is extremely dense, diffusion is the only mechanism necessary to convey the required volume of air to these muscles. However, in large

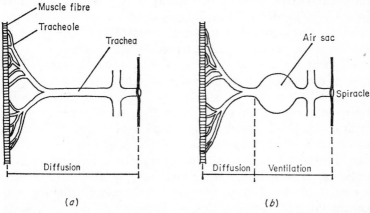

(a)   (b)

FIGURE 4.8 The effect of air sacs on the length of the diffusion path to the tissues. (a) without air sac; (b) with air sac. [b after Meglitsch (1967) *Invertebrate Zoology*. Oxford University Press: Oxford]

insects a ventilated air-sac system is necessary to shorten the diffusion path.

In some insects normal abdominal ventilatory pumping (chapter 8) is used and, in conjunction with the good connections between the abdominal and thoracic tracheal systems, it provides the pterothorax with sufficient air during flight (e.g. Hymenoptera). In others, however, abdominal pumping plays only a very minor role (e.g. Orthoptera and Odonata) and the thoracic air sacs are ventilated by movements of the thoracic sclerites directly associated with the wing movements (thoracic pumping). Thus in locusts, during average flight, the thorax receives about 320 l. air/kg/h. Abdominal pumping results in an exchange of about 180 l. air/kg/h and only 70 l. of this is pumped in and out of the pterothorax, leaving 250 l. to be provided by the thoracic pump. Of the remaining 110 l. produced by abdominal pumping 30 l. passes from the first spiracle posteriorly, avoiding the wing musculature, and 80 l. ventilates other parts of the animal. When necessary the capacity of the thoracic pump can be increased to at least three times the above value.*

In a few collembolans and the larvae of a few of the higher insects the spiracles are simple openings devoid of any closing mechanism. However, in most insects there is some mechanism for the restriction either of the spiracles or of the trachea at its junction with the atrium. In the Odonata for example the spiracle itself is closed by a special muscle which is sensitive to the local concentration of carbon dioxide, and opening is effected by the natural elasticity of the spiracular region. This type of mechanism is further elaborated in certain orthopteran spiracles where opening is aided by a specific opener muscle. Spiracular control in the Odonata and Orthoptera is dealt with in detail in chapter 8. An example of the type of closing mechanism which occurs at the junction of atrium and trachea can be found in the larvae of Lepidoptera. Here there is a cuticular bow that partly encircles the trachea and on the opposite side a rod. A muscle attaches these two structures and contraction of this muscle thus causes closure of the trachea. As in the Orthoptera opening is effected partly by the elasticity of the system and partly by the action of an antagonistic opener muscle.

In some insects the single spiracular opening is replaced by two or more openings. In certain beetle larvae there are two and these may either communicate via a common atrium or open directly into tracheae. Another, much more complex example is to be found in the third stage larvae of higher cyclorrhaphans (Diptera). Here there are two pairs of spiracles, one on the prothorax, the other towards the hind end of the abdomen, and these differ considerably from one another. Each prothoracic spiracle consists of several digitate processes; these open at their apices into a small atrium, which in turn opens into the main tracheal trunk. Each posterior spiracle on the other hand consists of a cuticular plate surrounded by a

* Data from Weis-Fogh (1967) *J. exp. Biol.*, **47**

peritreme. The plate normally bears three openings, traversed by fine cuticular rods (figure 4·9a), which lead into a common atrium lined with fibrous processes (figure 4·9b). Such elaborate systems are presumably an attempt to reduce water loss and to keep dust out of the tracheal system.

This control of respiration involving spiracular closing mechanisms and/or the possession of various filtering devices is termed 'Diffusion control'. There is another control mechanism, namely 'Ventilation control', which may be utilised

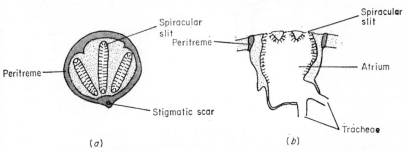

FIGURE 4·9 Posterior spiracle of *Calliphora erythrocephala*. (*a*) Surface view, (*b*) vertical section. [After Imms (1957) *A General Textbook of Entomology*. Methuen: London]

instead of (e.g. dragonfly larvae) or in conjunction with (e.g. locust) diffusion control.

Ventilation control of respiration (see chapter 8) is effected by movements of part of the abdomen and/or the thorax. In dragonflies (Odonata), locusts (Orthoptera) and cockroaches (Dictyoptera) dorsoventral flattening of the abdomen is involved; in others there is a telescoping of the abdominal segments (e.g. some Diptera and Hymenoptera), and in *Tipula* (Diptera) peristaltic waves pass over the abdomen.

Aquatic insects. Many aquatic insects have a closed tracheal system, that is to say the spiracles are all non-functional or absent, and oxygen diffuses into the tracheae either over the general body surface, which may be richly supplied with very fine tracheal branches, or via tracheal gills (chapter 2). However, a number do possess functional spiracles and these normally obtain their oxygen in one of four ways: by surface breathing, from aquatic plants, or by using temporary or permanent stores of air.

(a) *Surface breathing*. The basic problem for insects which utilise this method is one of penetrating the surface in order to bring the spiracle or spiracles into contact with the air, while at the same time avoiding flooding the tracheal system. This is achieved by having a hydrophobe area (i.e. one having a greater affinity for air than for water) around the spiracles, as in *Dytiscus* (Coleoptera) and certain dipteran larvae where

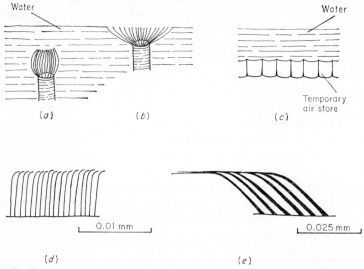

FIGURE 4·10 Spiracle surrounded by a ring of hydrophobe hairs (e.g. *Stratiomys* larva) in equilibrium below (a) and at (b) the surface of the water. (c) Temporary air store provided by short hydrophobe hairs (e.g. *Notonecta*). Plastron hairs of (d) *Aphelocheirus aestivalis* and (e) *Haemonia mutica*. [a–c after Wigglesworth (1953) *The Principles of Insect Physiology*. Methuen: London. d,e after Thorpe and Crisp (1949) *J. exp. Biol.*, **26**]

small glands at the spiracular openings produce an oily secretion. Alternatively the spiracles may be surrounded with semi-hydrofuge hairs which are hydrophobe on their inner surface, as in Notonecta (Hemiptera) and the larva of *Stratiomvs* (Diptera) (figure 4·10a, b).

In a number of dipteran larvae the spiracles are situated at the end of a siphon. This may be quite short (e.g. *Culex*) or long and retractile (e.g. *Eristalis*).

(*b*) *Use of aquatic plants.* The larvae of some coleopterans and dipterans obtain their oxygen from the air spaces of plants by inserting into them a specially modified respiratory siphon. There are also a few species that actually bite into the air spaces.

(*c*) *Temporary air stores.* The semi-hydrofuge, spiracular hairs of *Notonecta* can hold a very small reservoir of air when the animal submerges and this can be used as a temporary air store. The size of reservoir can be increased by having a large area of the body covered by hydrofuge hairs so that a thin film of air is retained. Thus, in *Notonecta* for example, the ventral surface of the body is covered with short erect hairs (figure 4·10*c*). Alternatively the space beneath the wing covers may be utilised, and dytiscid beetles carry a large bubble of air in this way.

These air stores have two principal functions. One is to act as a hydrostatic organ, the other to serve as a store of oxygen while the animal is submerged. Not only can oxygen be taken in through the spiracles from the air film, but a certain amount of oxygen can be extracted from the surrounding water by this layer. This is because the oxygen diffusion coefficient between water and air is more than three times that of nitrogen between the same media. Thus, if the oxygen tension in the water is higher than its partial pressure in the layer of air, equilibrium in the latter will tend to be restored by oxygen diffusing into it rather than by nitrogen diffusing out. This physical factor increases the efficiency of the air store about 13 times. Of course the nitrogen ultimately diffuses out and then the animal has to return to the surface, but if they are not actively swimming, animals can remain submerged for many hours with this type of air store.

The air film may be renewed in a variety of ways. The animal may simply push the hind end of its body through the surface film; some possess a respiratory siphon and yet others have hydrofuge hairs on the antennae which form an air channel along part of each antenna to the ventral body surface.

(*d*) *Permanent air stores: plastron respiration.* The plastron is a more-or-less permanent air store produced by the development of a semi-hydrofuge structure; it retains a constant volume of air in close proximity to the body and in contact with the spiracles when the animal is submerged. The water–air inter-face of this 'physical gill' is extensive. With very few exceptions a plastron is found only on species which inhabit a well-oxygenated habitat subject to rapid fluctuations in water level and which as a result frequently leave the animal exposed to the air. When the animal is submerged the retained layer of gas has a constant volume and so oxygen will diffuse out of the surrounding water into this layer in order to preserve the equilibrium—provided of course that the oxygen tension in the surrounding water is greater than the partial pressure of oxygen in the film of gas. Under well-oxygenated conditions the animal can live permanently submerged. Unfortunately, if the oxygen tension in the water is low this arrangement can work in the reverse direction equally well, removing oxygen from the insect. In times of drought the plastron ceases to function, except that near the spiracles it provides channels for the passage of respiratory gases, and in addition serves to restrict water loss through the spiracles.

The development of a plastron may be achieved by the further elaboration of hydrofuge hairs to retain the air film in such a way that it cannot be replaced by water. *Aphelocheirus* (Hemiptera) (figure 4·10d) and *Elmis* and *Haemonia* (Coleop-tera) (figure 4·10e) have a plastron of this type, most of the body bearing a dense mat of very fine hairs, each bent over at right angles at its tip and with the outer surface of its free end hydrophile. In *Aphelocheirus* there are approximately 2×10^6 hairs/mm². In this animal the tracheae open into an atrium from which a number of channels radiate, communicating with the surface via minute openings along their length. These channels and their openings are also lined with the plastron. In *Phytobius velatus* (Coleoptera) the plastron hairs are borne on scales which touch one another and, in some areas, overlap.

Alternatively the plastron may be a more rigid structure as in the case of that found on the spiracular gills of the pupae of certain dipterans and coleopterans (chapter 2). Here the plastron may be restricted to a series of rows (plastron lines) on the gill (figure 4·11), each consisting of a simple groove roofed over by a flat or an arched network, as in the tipulids *Lipsothrix remota* and *Antocha bifida* respectively (figure 4·12; plate II).

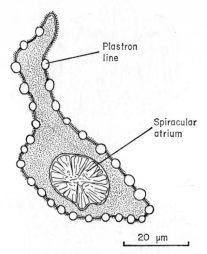

FIGURE 4·11 Transverse section through the 4th branch of a spiracular gill of *Antocha bifida* (see figure 2·10a). [After Hinton (1966)
Proc. R. ent. Soc. A, **41**]

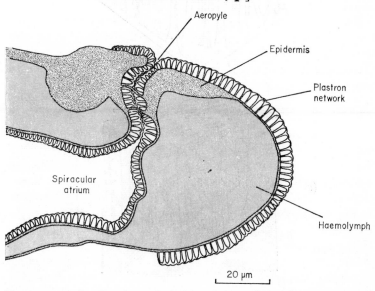

FIGURE 4·14 Section through the tip of a spiracular gill of *Geranomya
unicolor* to show the plastron network on the surface, and lining an
aeropyle and the spiracular atrium. [After Hinton (1968) *Advances
in Insect Physiol.,* **5**]

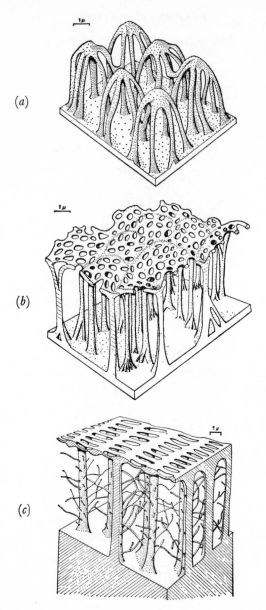

FIGURE 4·15 The plastron of (a) *Geranomya unicolor*, (b) *Dicranomyia trifilamentosa*, (c) *Aphrosylus celtiber*. (See Plate III.) [a,b from Hinton (1968) *Advances in Insect Physiology*, **5**; c from Hinton (1967) *J. mar. biol. Ass. U.K.*, **47**]

However, the plastron usually covers a large part or all of the gill. At its simplest it consists of a dense mat of small tubercles, as in the tipulid *Orimargula hintoni* (figure 4·13; plate II); but in most cases there is a series of vertical struts which branch at their tips to form an open hydrofuge network, and this usually lines the surface of the aeropyles (pores connecting the spiracular atrium with the surface) and spiracular atrium, as well as the surface of the gill (figure 4·14). The tipulids *Geranomya unicolor* and *Dicranomyia trifilamentosa* and the brachyceran *Aphrosylus celtiber* illustrate some of the variations in this pattern (figure 4·15 and plate III).

FIVE
Respiratory Pigments and the Transport of Respiratory Gases

In this chapter the occurrence and chemical nature of the respiratory pigments is dealt with, and this is followed by an account of their involvement in respiration and of the transport of oxygen and carbon dioxide.

There are four groups of respiratory pigments found in invertebrates—haemoglobin, chlorocruorin, haemocyanin and haemerythrin. Of these haemoglobin is the most widespread, occurring in many animals in a variety of phyla (table 5.1). Haemocyanin also occurs in a large number of animals, but is

TABLE 5.1. Distribution of respiratory pigments

(1) *Haemoglobin* (in plasma or cells) (red)	
Protostomia	
Acoelomata	
Platyhelminthes	Some parasitic rhabdocoels (e.g. *Syndesmis*) and trematodes
Nemertini	In plasma or cells
Pseudocoelomata	
Nematoda	In perivisceral fluid (e.g. *Ascaris*)
Coelomata	
Phoronida	In cells (e.g. *Phoronis*)
Mollusca	
Amphineura	Some. In plasma or cells
Gastropoda	Some prosobranchs and one pulmonate genus, *Planorbis*. In plasma
Bivalvia	A few (e.g. *Arca*). In cells
Echiuroidea	In cells in the coelomic fluid (e.g. *Urechis*)
Annelida	
Oligochaeta	In the plasma of all members of the class (e.g. *Lumbricus*)
Polychaeta	In the plasma in some (e.g. *Arenicola*), or in the cells in the coelomic fluid (e.g. *Glycerus*), or in both (e.g. *Terebella*)
Hirudinea	Some (e.g. *Hirudo*). In plasma
Arthropoda	
Insecta	A few. In the plasma in *Tendipes* and chironomid larvae. In special cells in larva of *Gastrophilus* and adults and larvae of the notonectids *Anisops* and *Buenoa*

TABLE 5.1 (continued)

Crustacea	In many lower crustaceans, but never in the Malacostraca. Branchiopods (e.g. *Artemia*, *Apus*, *Daphnia*), ostracods, copepods, some cirripedes and some branchiurans.
Deuterostomia	
Echinodermata	
Holothuria	In cells (e.g. *Cucumaria*)

---

(2) *Haemocyanin* (always in the plasma) (blue)
Protostomia
  Coelomata
    Mollusca

Amphineura	Some (e.g. *Cryptochiton*)
Gastropoda	Some, including all pulmonates except *Planorbis* (e.g. *Helix*)
Cephalopoda	All (e.g. *Loligo*)
Arthropoda	
Xiphosura	All (i.e. *Limulus*)
Crustacea	Stomatopods, decapods (e.g. *Homarus*), amphipods and isopods
Arachnida	Some (e.g. *Heterometrus*)

---

(3) *Haemerythrin* (always in cells) (violet)
Protostomia
  Pseudocoelomata

Nematoda	(e.g. *Ascaris*)
Coelomata	
Sipunculoidea	In both the coelomic fluid and the blood (e.g. *Sipunculus*
Annelida	
Polychaeta	Only in *Magelona*
Branchiopoda	In the coelomic fluid (e.g. *Lingula*)
Priapulida	Some

---

(4) *Chlorocruorin* (always in the plasma) (green)
Protostomia
  Coelomata
    Annelida

Polychaeta	Some, such as the serpulids and sabellids (e.g. *Sabella*). (Some serpulids, such as *Serpula*, have both chlorocruorin and haemoglobin in the plasma)

---

restricted to representatives of two phyla, the Arthropoda and the Mollusca. The other two pigments are very restricted in their distribution: haemerythrin is only found in the members of two small coelomate phyla, the Sipunculida and the Brachiopoda (lamp shells), in the pseydocoelomate phylum Priapulida and in one polychaete annelid (*Magelona*); chlorocruorin only occurs in a few polychaete families such as the Sabellidae and the Serpulidae. Even so some members of these families possess haemoglobin instead and *Serpula* is unusual in that it has both haemoglobin and chlorocruorin in its blood.

The respiratory pigments have in common an affinity for oxygen, with which they can combine reversibly. All four consist of a protein linked to a metal-containing prosthetic group. In haemoglobin and chlorocruorin this prosthetic group is an iron–porphyrin combination called a haem (i.e. an atom of iron linked to a porphyryl group), with the iron in the ferrous state. The prosthetic group of chlorocruorin (chlorocruorohaem) differs from that of haemoglobin (protohaem) in that a formyl chain replaces a vinyl chain on one pyrrole ring (position 2).

Chlorocruorohaem

M.....−CH_3
V......−$CH-CH_2$
P.....−$CH_2.CH_2.COOH$
F......−CHO

Protohaem

Iron is also present in the prosthetic group of haemerythrin, but the prosthetic group of haemocyanin contains copper instead. In both these pigments the metal is thought to be attached directly to the protein, and the prosthetic groups are not haems.

In haemoglobin and chlorocruorin, the minimum functional molecule which can combine with one molecule of oxygen contains a single atom of iron. In haemocyanin and haemerythrin the minimum functional molecule contains two atoms of metal. The molecular size of these pigments shows considerable variation in different animals and information about the size can be obtained from osmotic pressure measurements and by ultracentrifugation. From the latter a value known as the Sedimentation Constant (S) is obtained. This is defined as the sedimentation velocity in a unit centrifugal field at 20°C in a medium reduced to the viscosity of water. To avoid an unwieldy number of places after the decimal point this is normally expressed as cm $\times 10^{-13}$/sec/dyne. Larger molecules sediment out more quickly and hence the greater the sedimentation constant the larger the size of the molecule.

Haemoglobins consist of unit molecules. Each unit is composed of one haem together with its associated protein, globin, and has a molecular weight of between 17,000 and 18,000. In invertebrates the smallest known haemoglobins have two units (e.g. the larva of the midge *Chironomos plumosus* with a molecular weight of 31,400 and a sedimentation constant of 2.0), but range upwards in size and may be as large as several hundred units (e.g. *Arenicola marina* has a molecular weight of 3,000,000 and $S = 57.4$). (In vertebrate myoglobin and in the haemoglobin of *Lampetra* there is only a single unit of molecular weight 17,000; in *Myxine* haemoglobin there are two units, while in all other vertebrates (i.e. higher than the cyclostomes) the haemoglobin has four units, with a corresponding molecular weight of 68,000 and a value for S of about 4.4.)

Chlorocruorins consist of a large number of units and all have a molecular weight in the region of 3,000,000. The value of S thus shows little variation: in *Sabella* chlorocruorin it is 53; in that of *Serpula* it is 59.

Like the chlorocruorins, all haemocyanins are of large molecular size and so have a number of copper atoms in each molecule. In the Crustacea the molecular weight is either in the region of 400,000 with a sedimentation constant of about 16 (e.g. *Palinurus*) or about 800,000 with S equal to 24 (e.g. *Homarus*). The molecular weights of molluscan haemocyanins are considerably higher, ranging from 2,785,000 in the cephalopod *Octopus* $(S = 49.3)$ to 6,680,000 in the gastropod *Helix* $(S = 98.9)$.

In those haemerythrins for which data is available the molecular weight is always comparatively low, but there are still a number of iron atoms present in each molecule. In the sipunculids *Sipunculus* and *Golfingia* the molecular weights are 66,000 (12 iron atoms) and 119,400 (19 iron atoms) respectively.

In a number of animals the respiratory pigment is normally present in two sizes. In the crustacean *Daphnia* for instance the haemoglobin has a major component with a molecular weight of 422,000 which is stable over a range of pH from 4.5 to 10.5, and smaller amounts of a molecule with a molecular weight of 34,500. Outside the above range of pH the larger molecule dissociates into the smaller one. Other haemoglobins, such as that of *Arenicola* (figure 5·1*a*; plate IV), have only a single component. A similar situation occurs among the haemocyanins: in *Palinurus vulgaris* and *Panulirus interruptus* the haemocyanin has only one component, with a sedimentation constant of about 16 (16s component) over the normal physiological range of the blood pH. However, above a pH of about 9 (and below about pH 4 in *Panulirus* at least) this component can be dissociated into 4.1s and 6.4s components. On the other hand the haemocyanins of *Homarus vulgaris* and *Cancer pagurus* have two components in normal blood: a major 23s component and a 16*s* component in smaller amounts (figure 5·1*b*; plate IV). Again a dissociation can be elicited above a pH of 9, in this case into a 4s component. In *H. americanus* the 4s component has been demonstrated to be the minimum functional molecule of the haemocyanin, with a molecular weight of 68,700, and 12 of these subunits comprise the 24s component. The haemoglobin of *Arenicola* resembles this haemocyanin in that there are also 12 subunits in the largest component. In *Arenicola* they are arranged in two tiers of six (figure 5·1*a*) and the isolated subunits have an oxygen dissociation curve (p. 106) similar to that of whole molecules. Among the gastropod haemocyanins those of *Helix pomatia* (figure 5·1*c*; plate IV) and *Buccinum undatum* have two normal components, a major one of about 100s and a smaller one of about 63s. Again at about pH 9 dissociation occurs. In *Neptunea antigua* the 63s component is absent. Other haemocyanins, such as those of cephalopods (figure 5·1*d*; plate IV), arachnids and *Limulus*, have more than one component, the largest of which are 62s, 38s and 56s respectively.

Respiratory pigments with a molecular weight over about 200,000 are always found in solution in the blood, never inside the cells. This includes the chlorocruorins, haemocyanins and most haemoglobins, and is the more normal situation in invertebrates. However, there are some invertebrates which possess pigments of lower molecular weight (haemerythrins and some haemoglobins) and, with the exception of those in two insects, *Tendipes* and *Chironomus* larvae, they are all to be found inside blood cells and are always associated with a degenerate or non-existent circulatory system in which the respiratory pigment is in the coelom. The retention of pigments of small molecular weight inside corpuscles no doubt prevents them from being removed by the excretory system. It is known that proteins of low molecular weight can be removed from the blood by a process of ultrafiltration and it may be significant that *Tendipes* and *Chironomus* are both insects, a group in which the excretory system does not depend on such a process.

The pigments have characteristic colours which depend on their state of oxygenation. Haemoglobin for example is purple, but turns red in its oxygenated form (oxyhaemoglobin); haemocyanin is colourless, but becomes blue when oxygenated. Furthermore, these pigments have characteristic absorption spectra. Oxyhaemoglobin has three main bands, one in the yellow (alpha band), one in the green (beta band) and one in the violet (Soret band); in deoxygenated haemoglobin the alpha and beta bands are replaced by a single broad band in the yellow–green and the Soret band has a slightly higher wavelength. Thus the absorption spectra can be used to give some indication of the proportion of pigment in an oxygenated condition.

The respiratory organs discussed in the three previous chapters have in common the function of facilitating the exchange of respiratory gases, either by bringing the fluid of the circulatory system into close proximity with the environment or by bringing the environment into the close proximity of the tissues. The maximum partial pressure or tension of oxygen (pO_2) in the environment rarely exceeds 159 mm mercury (dry air at sea level) and under such advantageous conditions the dissolved oxygen in the blood in equilibrium with this environment may reach 0.5 ml per 100 ml blood (0.5 volumes per cent). However, the blood of invertebrates may contain almost 10 vol per cent and this is made possible by the properties of the respiratory pigments.

The oxygen capacity of the blood can only be increased by

an increase in the amount of respiratory pigment and two antagonistic factors are important in this respect. If the respiratory pigment has a high molecular weight the viscosity of the blood may become so great as to prevent it from flowing through the vessels. However, reduction of the molecular weight leads to an increase in osmotic pressure of the blood, thus affecting the water balance between it and the tissues, and so the pigment must be contained within cells to prevent it from contributing to the osmotic pressure. This is largely a theoretical

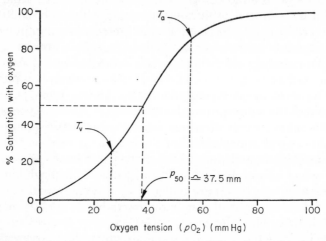

FIGURE 5·2 An oxygen dissociation curve. The relationship between the percentage saturation of the blood with oxygen and the oxygen tension (pO_2) in mm Hg. The half-saturation tension (p_{50}) is 37.5 mm. T_a is the oxygen tension in the blood arriving at the tissues; T_v that in the blood arriving at the respiratory organs.

argument as far as invertebrates are concerned since the highest oxygen capacities ever recorded (9.9 vol per cent) is in the annelid *Arenicola*, in which the pigment has a molecular weight of 3,000,000 and is freely dispersed in the plasma. This high molecular weight keeps the blood osmotic pressure low and it is presumably not so concentrated as to significantly retard the blood flow. These considerations become more important in vertebrates, some of which have oxygen capacities exceeding 20 vol per cent, and in which the molecular weight of the pigment is always comparatively small (about 68,000).

Under normal conditions the amount of dissolved oxygen increases directly with an increase in oxygen tension (pO_2) of the environment. The presence of a respiratory pigment not only causes the percentage saturation with oxygen to rise more steeply with increase in environmental pO_2, but also affects the shape of this absorption curve (usually referred to from the reverse point of view as a dissociation curve) (figure 5·2). This normally tends towards a sigmoid shape, but may be almost hyperbolic. Ideally the oxygen tension of the blood in the respiratory organ (T_a) and that in the tissues (T_v) should be such that they lie adjacent to the steep section of the dissociation curve. The oxygen tension required to saturate 50 per cent of the pigment (p_{50}) is used to describe the oxygen affinity of the pigment and this value gives some indication of the effective slope of the dissociation curve. It follows that the oxygen affinity of the pigment is low when p_{50} is high, since this indicates that the pigment releases the bound oxygen at high values of pO_2. Conversely the oxygen affinity is high when p_{50} is low.

The way in which oxygen is bound to the respiratory pigment normally involves more than a simple reaction between the two. The pigment exists as a molecule which is composed of at least two minimum functional units; in *Panulirus* for example the normal molecule of haemocyanin is composed of six units, each with its two atoms of copper. If each oxygen-binding site were independent of the others and the reaction at each of them could be written simply as

$$Cu_2 + O_2 = Cu_2O_2$$

then a hyperbolic dissociation curve would result. The sigmoidal nature of the curve generally obtained in practice is due to the interaction between the oxygen-binding sites, with possibly each reaction having its own specific rate. Thus

$$Cu_{12} + O_2 = Cu_{12}O_2$$
$$Cu_{12}O_2 + O_2 = Cu_{12}O_4$$

$$\cdot$$
$$\cdot$$
$$\cdot$$

$$Cu_{12}O_{10} + O_2 = Cu_{12}O_{12}$$

The dissociation curve is approximated over its middle portion by Hill's empirical equation

$$\bar{Y} = \frac{Lp^n}{1 + Lp^n} \qquad (1)$$

where $\bar{Y}$ = the fraction of pigment combined with oxygen; $p = pO_2$ in mm mercury; L = the equilibrium coefficient between oxygenated and deoxygenated pigment; n is a measure of the interaction of the sites and equals 1 when the curve is a true hyperbola. Equation (1) can be rewritten

$$Y = \frac{100Lp^n}{1 + Lp^n} \qquad (2)$$

where Y = the percentage saturation of the pigment with oxygen. Equation (2) can be simplified by dividing both sides by $100 - Y$. Thus

$$\frac{Y}{100 - Y} = \frac{\dfrac{100Lp^n}{1 + Lp^n}}{100 - Y} \qquad (3)$$

where $\dfrac{Y}{100 - Y}$ is the ratio between saturated and unsaturated pigment.

From equation (2)

$$100 - Y = 100 - \frac{100Lp^n}{1 + Lp^n}$$

$$= \frac{100 + 100Lp^n - 100Lp^n}{1 + Lp^n}$$

$$= \frac{100}{1 + Lp^n}$$

Substituting for $100 - Y$ in the right hand side of equation (3) gives

$$\frac{Y}{100 - Y} = \frac{\dfrac{100Lp^n}{1 + Lp^n}}{\dfrac{100}{1 + Lp^n}}$$

$$= \frac{100Lp^n}{1 + Lp^n} \cdot \frac{1 + Lp^n}{100}$$

$$= Lp^n \qquad (4)$$

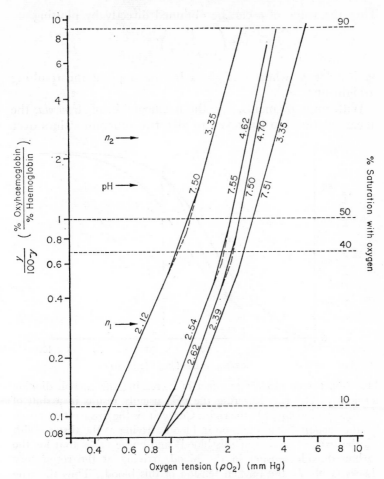

FIGURE 5·3 *Arenicola marina*. Logarithmic plot of $Y/(100 - Y)$ (% Oxyhaemoglobin/% Haemoglobin) against oxygen tension to show the diphasic haem–haem interactions between 10 per cent and 90 per cent saturation with oxygen. Dashed lines indicate the slopes of the two main phases. [After Weber (1970) *Comp. Biochem. Physiol.*, **35**]

From equation (4)

$$\log \frac{Y}{100 - Y} = \log L + n \log p \qquad (5)$$

Thus the value of n can be obtained directly by plotting

$$\log \left(\frac{Y}{100 - Y} \right)$$

against $\log p$, when it is given by the slope of the resulting straight line.

With some pigments, e.g. the haemoglobin of *Arenicola*, the linear relationship is broken to give two different values over

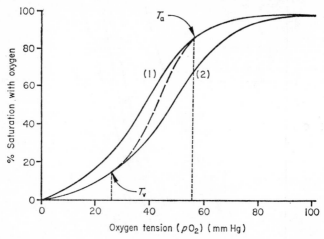

Oxygen tension (pO_2) (mm Hg)

FIGURE 5·4 *Bohr effect.* The effect of increasing the carbon dioxide tension (pCO_2) (i.e. lowering the pH) usually results in a shift of the dissociation curve to the right. Curve (1) is the dissociation curve for the carbon dioxide tension in blood arriving at the tissues (with T_a the oxygen tension of this blood) and curve (2) is that for the carbon dioxide tension in the blood arriving at the respiratory tissues (with T_v the oxygen tension of this blood). Thus the true dissociation curve is represented by the dashed line between T_a and T_v. Note that this is steeper than curves (1) and (2).

different parts of the pO_2 range and this phenomenon is associated with a sigmoidal rather than a hyperbolic dissociation curve. In the haemoglobin of *Arenicola* between 10 per cent and 90 per cent oxygen saturation there are two separate slopes (n_1 and n_2) (figure 5·3) with the change-over at about 40 per cent. n_1 is lower than n_2, indicating a weaker haem–haem interaction at low oxygen tensions. The higher the value for n the more haems are involved in the interactions. Below about

10 per cent oxygen saturation n tends to approach unity, which probably reflects the low probability of more than one oxygen molecule being bound to the same pigment molecule at low saturation values.

There are two factors of considerable importance which may have an effect on the parameters of the dissociation curve. These are the acidity of the blood and the temperature.

5.1. Effect of acidity (Bohr effect) In general if the acidity of the blood increases (i.e. the pH decreases) its affinity for

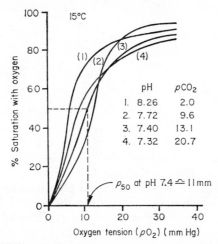

FIGURE 5·5 *Helix pomatia:* haemocyanin. An example of a dissociation curve which is hyperbolic at low CO_2 tensions. Increase in CO_2 tension tends to make the curve more sigmoidal. [After Spoek, Bakker and Wolvekamp (1964) *Comp. Biochem. Physiol.*, **12**]

oxygen decreases and this has the overall effect of moving the dissociation curve to the right (figure 5·4). The acidity is affected by the carbon dioxide tension (pCO_2) of the blood. In the region of the active tissues the pCO_2 tends to be high, thereby increasing the acidity of the blood and making more of the bound oxygen available to the tissues. In other words the blood releases more oxygen at a given pO_2 if the pCO_2 is high. Conversely, in the respiratory organ carbon dioxide is lost to the environment, so the blood acidity decreases and this enables the respiratory pigment to take up a maximal amount of oxygen. The true dissociation curve is thus one lying between the two curves of figure 5·4, joining the left-hand curve at T_a and the right-hand curve at T_v. As can be seen the effective part of the curve is steeper because of this *Bohr effect*, and the greater the effect the steeper will it become. In a number of animals, such as the gastropod molluscs *Helix*, *Planorbis* and *Agriolimax* and the lobster *Homarus*, the dissociation curve is almost hyperbolic in the absence of carbon dioxide (or at low

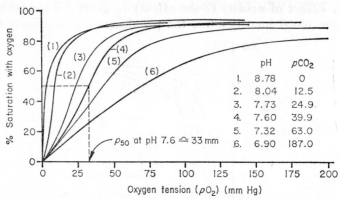

FIGURE 5·6 *Root effect. Agriolimax:* haemocyanin. The saturation level of this pigment becomes lower at high tensions of CO_2, and this is known as the Root effect. [After Prosser and Brown (1961) *Comparative Animal Physiology*. Saunders: Philadelphia]

values of pCO_2). An additional effect of increase in pCO_2 in these animals is to make the curves more sigmoidal (figure 5·5), and so it appears that increase in acidity increases the inter-action between oxygen-binding sites. Finally, in some animals the shift of the equilibrium curve to the right is accompanied by flattening of the curve at below complete saturation (e.g. *Agriolimax*), and this is called the *Root effect* (figure 5·6).

There is a limiting value of pH (about 6.2 for example in *Homarus*) below which the Bohr effect is reversed. This is normally outside the physiological range of the animal, but in some cases such as in *Limulus* (Xiphosura), *Busycon* (Mollusca, Gastropoda) and *Heterometrus* (Arachnida, Scorpionidea) there is a reversal within the normal range (figure 5·7).

5.2. Effect of temperature With increase in temperature the percentage saturation of the pigment with oxygen decreases for a given environmental pO_2 (figure 5·8). As in the case of

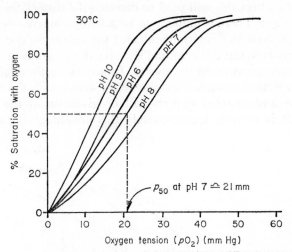

FIGURE 5·7 *Heterometrus pulvipes:* haemocyanin. This pigment shows a reversed Bohr effect within the physiological range of the animal. [After Padmanabhanaidu (1966) *Comp. Biochem. Physiol.*, **17**]

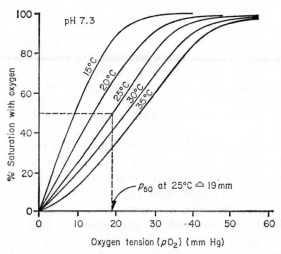

FIGURE 5·8 *Heterometrus pulvipes:* haemocyanin. The effect of increase in temperature is to shift the dissociation curve to the right. [After Padmanabhanaidu (1966) *Comp. Biochem. Physiol.*, **17**]

the Bohr effect this will tend to displace the dissociation curve to the right. There is unlikely to be a large temperature difference between the environment and the tissues in the case of invertebrates, but if the animal is active there may be a slightly higher temperature in the tissues. In any event the higher the environmental temperature the more active will the animal become and so more oxygen will be given up at the tissues when it is needed. Unfortunately this advantage is partially

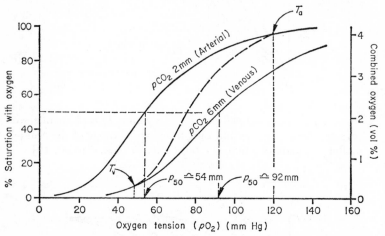

FIGURE 5·9 *Loligo peali:* haemocyanin. This is a low oxygen affinity pigment. The O_2 tension in the post-branchial vessel (T_a) is 120 mm ($pCO_2 = 2$ mm); that in the posterior venous sinuses (T_v) 48 mm ($pCO_2 = 6$ mm). The true dissociation curve is represented by the dotted line between T_a and T_v. [After Jones (1963) *Problems in Biology*, **1** (Ed. Kerkut). Pergamon: Oxford]

offset by some reduction in the uptake of oxygen from the environment. In some animals (e.g. *Limulus, Palinurus* and *Heterometrus*) increase in temperature progressively changes the shape of the curve from near hyperbolic to sigmoid and thus increases the interaction between oxygen-binding sites.

5·3. Chlorocruorin and haemocyanin It may be thought that in an environment rich in oxygen, such as the sublittoral zone, the respiratory pigment would have a high oxygen capacity and release most of its oxygen at fairly high values of pO_2 (i.e. have a low oxygen affinity); and indeed this is the case in, for example, the polychaete *Sabella* (chlorocruorin) and the cephalopod mollusc *Loligo* (haemocyanin), both of which are marine animals. In the former the oxygen capacity has been estimated at 9.1 vol per cent (one of the highest recorded in an invertebrate); in *Loligo* at between 3.4 and 4.3 vol per cent. The value of p_{50} in *Sabella* is 8 mm mercury at pH 8.0 and 10°C, and 29 mm at pH 7.4 and 26°C; in *Loligo* (figure 5·9) it is 36 mm in the absence of carbon dioxide and rises to 96 mm at a pCO_2 of 6.5 mm, which is about the level in the venous blood. In both cases the Bohr effect is normal and quite large, and so plays an important part in the release of oxygen to the tissues and its uptake from the environment. Thus, in *Loligo* the small decrease in pH in the tissue capillaries caused by the increase in pCO_2 from 2 mm to 6 mm is sufficient on its own to release between one quarter and one third of the total bound oxygen. The temperature effect is also normal. Although the oxygen affinity in *Sabella* is not so low as that in *Loligo* this is compensated for by an oxygen capacity over double that of *Loligo* (figure 5·9 and table 5.2*a*).

The decapod crustaceans which, like *Loligo*, possess haemocyanin present a rather anomalous picture. Thus although many such as the lobster *Panulirus* live in an oxygen-rich environment similar to that of *Loligo*, the oxygen capacity of the blood is lower than that of the cephalopods, and yet the oxygen affinity is rather higher than in *Sabella*. For example, in *Panulirus interruptus* (figure 5·10) the oxygen capacity is only 1.0 to 2.8 vol per cent and the value of p_{50} varies from 6 mm in the absence of carbon dioxide at 15°C to a mere 19 mm at a pCO_2 of 32 mm at 23°C. At first sight it appears as though the respiratory pigment could only increase the oxygen affinity of the blood five or six times at most, since one would expect the environment to be almost saturated with oxygen and this itself would cause around 0.5 vol per cent to be present in the blood. However, the absorption process across the gills is apparently not very efficient and full utilisation of the oxygen tension of the environment is not made. The net effect is that the arterial blood is only 54 per cent saturated in animals in well-aerated sea water, with a resultant oxygen content (as distinct from oxygen capacity) of only 0.82 vol per cent. This gives a blood pO_2 of 7 mm at which tension the dissolved oxygen will

TABLE 5.2. Low oxygen affinity pigments

	Oxygen affinity				Oxygen content			Oxygen capacity	Bohr effect	Temp. effect
	p_{50} (mm Hg)	pH	pCO_2 (mm Hg)	Temp. (°C)	Content (vol %)	pO_2 (mm Hg)	% Saturation	(vol %)		
(a) For transport from an environment of high pO_2										
Sabella spallanzanii (C)	9	8		10						Normal
Loligo pealei (Hc)	26.5 / 36	7.7	0	20	{ A.4.27 / V.0.37 }	120 / 30	97 / 8	9.1	Normal	Normal
Octopus vulgaris (Hc)	c54 / c92		2 / 6	23 / 23	{ A.4.64 / V.0.40 }	84* / 1*	100* / 9*	3.4–4.1	Normal	
Panulirus interruptus (Hc)	6.5 / 9.2	7.53 / 7.53		15 / 20	{ A.0.82 / V.0.35 }	7 / 3	54 / 22	1.0–2.8	Normal	Normal
Loxorhynchus grandis (Hc)	5.4 / 15.5	7.52 / 7.52		15 / 20	{ A.0.41 / V.0.17 }	8 / 3	68 / 30	0.9–1.5		Normal
Homarus americanus (Hc)					{ A.0.44 / V.0.18 }	5 / 2	49 / 20	1.2–1.4		
Cancer irroratus (Hc)	12		0	23				0.1–1.2		
(b) For transport from an environment with extremes of pO_2										
Busycon canaliculatum (Hc)	12		0	23	{ A.1.65* / V.0.33* }	36 / 6	95* / 20*	1.7–2.9	Normal and Reversed	
Helix pomatia (Hc)	7 / 13 / 11 / 8.5	8.3 / 7.7 / 7.4 / 7.3	2.0 / 9.6 / 13.1 / 20.7	15 / 15 / 15 / 15					Normal and Reversed	Normal
Limulus polyphemus (Hc)	11		0	23				0.1–1.7		
Heterometrus pulvipes (Hc)	16.5	7.3		30				1.8	Normal and Reversed	Normal

(c) For transport from an environment when the pO₂ becomes low, and with an additional storage function at low pO

Nephthys hombergii (Hb)

								Storage function
(i) Vascular	5.0	7.0		15				
	5.5	7.4		15				
	5.0	7.0		15				
	7.5	7.4		15				
(ii) Coelomic	12		15		12			
Urechis caupo (Hb)	12		22		c.2.9–4.1†	97†	2.7–7.2	Normal
Sipunculus nudus (He)	37	8	34	19	32	1.1	90	1.0	Normal
Dendrostomum zostericolum (He)								
(i) Vascular	15	6.67–7.25	20					None
(ii) Coelomic		5.20–7.86	20					None

					Storage function
(d) A storage function only					
Anisops pelluceus (Hb)	28	—	24		None

A, Postbranchial; V, Prebranchial; C, Coelomic
* Calculated by Redmond
† When the environmental pO₂ is high

Data from Fox (1932) *Proc. Roy. Soc. Lond. B,* **111** (*Sabella*); Florkin (1934) *Ann. Physiol. Physicochem. biol.,* **32** (*Nephthys*); Jones (1955) *J. exp. Biol.,* **32** (*Nephthys*); **10** (*Sipunculus*); Padmanab-hanaidu (1966) *Comp. Biochem. Physiol,* **17** (*Heterometrus*); Redfield, Coolidge and Hird (1926) *J. Biol. Chem.,* **69** (*Loligo, Busycon, Limulus* and *Cancer*); Redfield and Florkin (1931) *Biol. Bull, Woods Hole,* **61** (*Urechis*); Redfield and Goodkind (1929) *Brit. J. exp. Biol.,* **6** (*Loligo*); Redmond (1955) *J. cell. comp. Physiol.,* **46** (*Panulirus, Loxorhynchus* and *Homarus*); Spoek (1964) *Comp. Biochem. Physiol.,* **12** (*Helix*); and Winterstein (1909) *Biochem. Zeit.,* **19** (*Octopus*); Miller (1966) *J. exp. Biol.* **44** (*Anisops*) only be 0.03 vol per cent. Hence the presence of haemocyanin enables the arterial blood to contain about 27 times that present in solution alone. The temperature effect is normal and of reasonable magnitude, but the Bohr effect is of negligible importance since there is only about 0.4 mm difference between the tension of carbon dioxide in the arterial blood and that in the venous blood (figure 5·10 and table 5.2a).

Most gastropod molluscs also possess haemocyanin; the marine ones (e.g. *Busycon*) are of particular interest in that

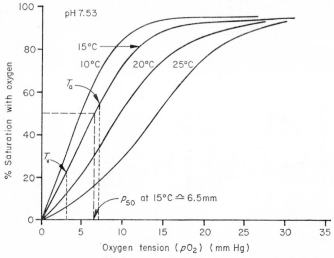

FIGURE 5·10 *Panulirus interruptus:* haemocyanin. This is a high oxygen affinity pigment. The oxygen tension in the pericardium (T_a) is 7 mm; that in the abdominal sinus (T_v) 3 mm. [After Redmond (1955) *J. Cell. comp. Physiol.*, **46**]

(with the possible exception of the conch *Pleuroploca*) the Bohr effect is reversed within their physiological range. This reversal also occurs in the terrestrial shelled snails (e.g. *Helix*) and in certain arthropods such as the king-crab *Limulus* and the scorpion *Heterometrus fulvipes* (figure 5·7). In terrestrial gastropods without shells (e.g. *Agriolimax columbianus*) the Bohr effect is normal. The oxygen capacity of the pigment in all of these animals is similar to that of the decapod crustaceans: in *Busycon* it is 2.3 vol per cent and in *Heterometrus* about 2 vol per cent. However, the oxygen affinity is lower than that of

PLATE IV

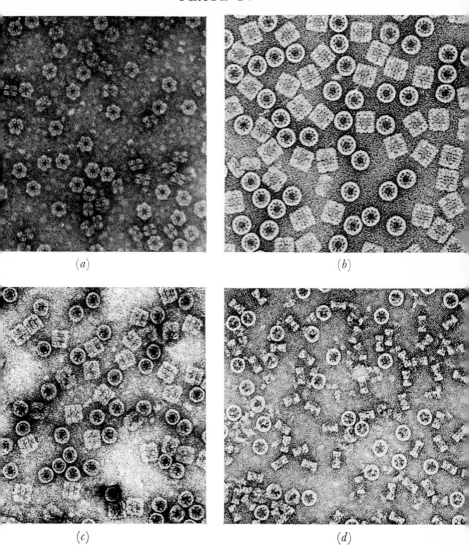

FIGURE 5·1 Electron micrographs of respiratory pigment molecules.
(a) Haemoglobin of *Arenicola marina* negatively stained with uranyl
oxalate ($\times$ 150,000); (b) haemocyanin of *Helix pomatia*, at pH 7.6,
negatively stained with uranyl EDTA ($\times$ 150,000); (c) haemocyanin of *Neptunea antigua* negatively stained with uranyl EDTA
($\times$ 135,000); (d) haemocyanin of *Octopus vulgaris*, at pH 7.8,
negatively stained with uranyl oxalate ($\times$ 140,000). [a by courtesy
of Weber and van Bruggen; b–d courtesy of van Bruggen. b,d from
van Bruggen, Wiebenga and Gruber (1962) *J. mol. Biol.*, **4**; c from
van Bruggen, Schuiten, Wiebenga and Gruber (1963) *J. mol. Biol.*, **7**]

PLATE V

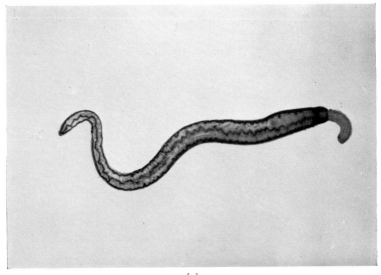

(a)

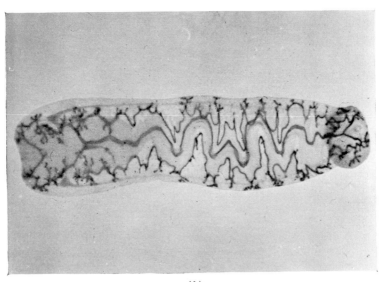

(b)

FIGURE 6·1 The circulatory system of the nemertines (a) *Amphiporus lactifloreus* and (b) *Malacobdella grossa*, as demonstrated by the leucine aminopeptidose technique. [From Gibson and Jennings (1967) *Comp. Biochem. Physiol.*, **23**]

PLATE VI

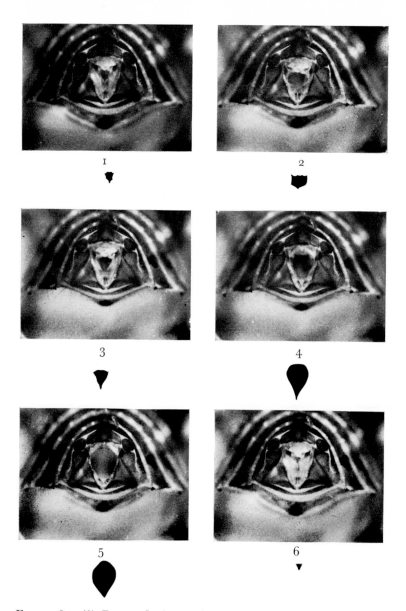

FIGURE 8·3 (b) Dragonfly larva. Selected views of the anal valve. The numbers correspond to the positions shown in figure 8·3 (a); the black shape below each photograph indicates the open anal valve area in that picture. [After Mill and Pickard (1972) J. exp. Biol., **56**]

decapods, with values of p_{50} intermediate between those of *Panulirus* and *Loligo*, and they are more like the latter in that the arterial blood is almost fully saturated with oxygen (table 5.2*b*). In the Polyplacophora (chitons) the pigment is also fairly well saturated with oxygen (about 90 per cent) and the oxygen affinity is low. The Bohr effect in these animals is small and, because of the abundant supply of oxygen in most cases, is not likely to play any major role in respiratory exchange. *Ischnochiton conspicuous* has a higher oxygen affinity than the other chitons so far studied and this probably relates to its habitat, where oxygen tensions may often be low.

It is generally considered that the reversed Bohr effect is an adaptation to an environment in which the oxygen tension may at times become low with a concurrent rise in the amount of carbon dioxide. Under these conditions a shift of the curve to the left would enable the arterial blood to maintain a high concentration of oxygen. In the shelled gastropods the only time when such conditions could occur are when the animal is withdrawn into its shell. It may then be a means either of extracting oxygen from the water or air retained in the mantle cavity, or of attracting the oxygen from the blood to the more active tissues. The latter could only work if the circulation is poor, which is probably the case in these animals when they are not making any body movements. *Busycon* differs from *Helix*, *Limulus* and *Heterometrus* in that its Bohr effect is reversed over the entire physiological range, whereas in the others the effect is at first normal with increase in pCO_2 and becomes reversed after a critical tension is reached (about pH 8 in the case of *Heterometrus*) (table 5.2*b*).

5.4. Haemoglobin We will now turn our attention to the haemoglobins. Unlike haemocyanins and chlorocruorins, the oxygen affinity of haemoglobins is generally high, although there are a few exceptions. The earthworms *Lumbricus* and *Allolobophora* are examples of animals with high oxygen-affinity haemoglobin (figure 5·11). In the former the value of p_{50} varies between about 2 and 8 mm, and is even lower in *Allolobophora*; thus the blood of these animals has a fairly constant oxygen uptake over a wide range of environmental oxygen tensions. In *Lumbricus* the haemoglobin makes a major contribution to this uptake at least down to an environmental pO_2

E

of 40 mm. This type of system must reflect the very low oxygen tensions in the venous blood which are necessary to allow the pigment to unload its oxygen to the tissues. The Bohr and temperature effects are normal. It has been suggested that although *Lumbricus* normally lives in an environment of high pO_2 this mechanism is necessitated by the absence of respiratory

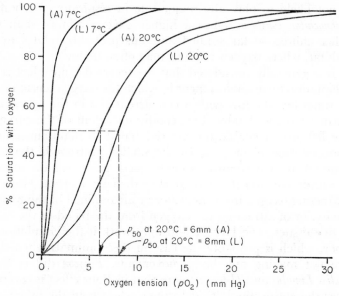

FIGURE 5·11 *Allolobophora terrestris* (*A*) and *Lumbricus terrestris* (L): haemoglobin. These are high oxygen affinity pigments. [After Haughton, Kerkut and Munday (1958) *J. exp. Biol.*, **35**]

organs and a high resistance to diffusion over the body surface (table 5.3*a*).

A similar situation exists in some other annelids. In the aquatic oligochaete *Tubifex* for example the oxygen affinity is very high, with a value for p_{50} of only 0.6 at 17 °C in the absence of carbon dioxide. This high affinity pigment is presumably needed because of the wide variation in oxygen tension experienced in the environment. The venous oxygen tension will be kept at a low level by the activity of the tissues (table 5.3*b*).

Haemoglobin also occurs in the blood of some lower crustaceans such as *Daphnia*. This is an unusual animal in that it

TABLE 5.3. High oxygen affinity pigments

	Oxygen affinity				Oxygen content (vol %)	Oxygen capacity (vol %)	Bohr effect	Temp. effect
	p_{50} (mm Hg)	Acidity		Temp. (°C)				
		pH	pCO₂ (mm Hg)					
(a) For transport from an environment of normally high pO₂ without specialised respiratory structures								
Lumbricus terrestris (Hb)	2 8	— —		7 20			Normal (moderate)	Normal
Allolobophora longa (Hb)	3.5 4.8 0.7 6	7.72 7.21 — —		7 20				
(b) For transport from an environment with extremes of pO₂								
Tubifex sp. (Hb)	0.5 0.6		0	10				
Daphnia magna (Hb)	2.0 4.9		0 (1%)	17 17			Normal (small)	Normal (small)
(c) For transport from an environment with extremes of pO₂, and with an additional storage function at low pO₂								
Arenicola marina (Hb)	1.5 2.6 1.7 3.7	7.8 6.9	0 (1%)	10 17 20 20	C2.08*–6.38*	9.7	Normal (small)	Normal (small)
Chironomus riparius (Hb)	5 0.5 0.6	5.4–6.8	(1%)	20 10 17			Absent	Normal (v. small)
(d) Only for transport from a pulmonary store								
Planorbis corneus (Hb)	1.5 2.9 3	8.5–8.6	0 (1%) 1–4	10 17 19–20		1.4–2.9	Normal and Reversed (small)	Normal (small)

C, Coelomic
* Calculated from Barcroft and Barcroft

Data from Barcroft and Barcroft (1924) Proc. Roy. Soc. Lond. B., 96 (Arenicola); Borden (1931) J. mar. biol. Ass., 12 (Arenicola and Planorbis); Fox (1945) J. exp. Biol., 21 (Tubifex, Daphnia, Arenicola, Chironomus and Planorbis); Haughton, Kerkut and Monday (1958) J. exp. Biol., 35 (Lumbricus and Allolobophora); Jones (1964) Comp. Biochem. Physiol., 12 (Planorbis); and Manwell in Prosser and Brown (1961) Comparative Animal Physiology, Saunders: Philadelphia (Arenicola) can rapidly increase the concentration of pigment in its blood when the oxygen tension of the environment falls by only a small amount, which indicates that the pigment is functional at high oxygen tensions. It is difficult to obtain accurate data from small animals with only a limited amount of blood, which is usually the case in those possessing haemoglobin. One technique is to measure the absorption spectrum of the pigment and use this to determine the value of p_{50}, since haemoglobin and oxyhaemoglobin have different absorption spectra (p. 105). Another technique is to make use of the fact that carbon monoxide combines with haemoglobin to form carboxyhaemoglobin, the dissociation curve obtained in the presence of carbon monoxide indicating the proportion of oxygen bound by the pigment. The difficulties of obtaining direct measurements may account for the observation that in *Daphnia* the oxyhaemoglobin appears to be completely dissociated when the external oxygen tension has only fallen to 28 mm, in spite of the fact that the value of p_{50} is apparently considerably lower. In any event a high gradient across the respiratory surface is indicated, as in *Lumbricus* (table 5.3*b*).

At first sight the respiratory pigment in a number of other haemoglobin-bearing animals, such as the marine oligochaete *Arenicola marina*, the aquatic larva of the midge *Chironomus plumosus*, and the gastropod mollusc *Planorbis corneus* (the only gastropod genus possessing haemoglobin), may be thought to function in a similar manner. They all have, for example, a high affinity for oxygen. However, a storage function has been postulated for the pigment in these cases.

Planorbis corneus is aquatic and has a secondary gill, while retaining an air-filled lung. Even in well-oxygenated water about a third of the total uptake of oxygen is pulmonary and this increases as the pO_2 of the water decreases. The haemoglobin of this animal has a p_{50} of about 3 mm at 20°C in the presence of 1–4 mm of carbon dioxide (figure 5·12 and table 5.3*d*). At low tensions of carbon dioxide the Bohr effect is normal, but a reversal occurs above about 20 mm. It is unusual for a haemoglobin to show a reversal within the physiological range of the animal, although haemocyanin-bearing gastropods such as *Busycon* and *Helix* exhibit this phenomenon. However, in its normal habitat (water without much decaying organic matter) the pCO_2 rarely exceeds 15 mm. Thus, except when the snail withdraws into its shell, the carbon dioxide tension in the blood will keep fairly low, since carbon dioxide readily diffuses across the animal's skin. The temperature effect is normal and the oxygen capacity is fairly low (0.94–2.94 vol per cent), but is presumably compensated for by the comparatively large volume of blood in this animal. The oxygen content of the habitat is very variable and values ranging from

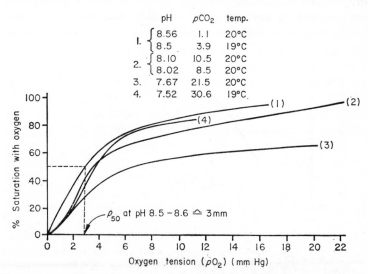

	pH	pCO_2	temp.
1.	8.56	1.1	20°C
	8.5	3.9	19°C
2.	8.10	10.5	20°C
	8.02	8.5	20°C
3.	7.67	21.5	20°C
4.	7.52	30.6	19°C

FIGURE 5·12 *Planorbis corneus:* haemoglobin. This haemoglobin is unusual in that it shows a reversed Bohr effect within the physiological range of the animal. [After Zaaijer and Wolvekamp (1958) *Acta Physiol. Pharmacol. Neerl.*, **7**]

10 to 490 mm have been recorded, the latter indicating a very high level of supersaturation. Although it might be thought useful for the pigment to store oxygen in *Planorbis*, this is unlikely, since the effect of any low tensions of oxygen in the water is partially negated by the air breathing. During a dive the oxygen tension in the lung is reduced from about 125 mm to 21 mm, and the evidence indicates that the haemoglobin will only start to transport oxygen when the oxygen tension in the lung falls below 60 mm, that is in the later stages of the dive. Thus the function of the pigment seems to be to facilitate

the exploitation of the pulmonary oxygen store and it apparently plays little part in the transport of oxygen from the water.

Arenicola marina has a value for p_{50} in the order of 2.0 to 8.3 mm. Its oxygen capacity is between 2.0 and 9.9 vol per cent, the latter figure being the highest recorded from any invertebrate (table 5.3*c* and figure 5.13). This animal lives in a U-shaped tube in the sand which it periodically irrigates with well-aerated sea water. However, this irrigation can be interrupted in two ways: firstly, when the sand is exposed for long periods at low water it is thought that the animal may resort in part to aerial respiration by drawing bubbles of air

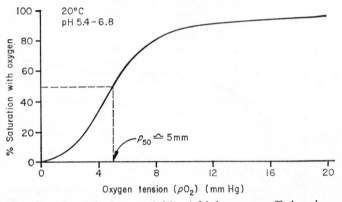

FIGURE 5·13 *Arenicola:* haemoglobin. A high oxygen affinity pigment with no noticeable Bohr effect. [After Prosser and Brown (1961) *Comparative Animal Physiology*. Saunders: Philadelphia]

down over its gills; secondly, while the tube is still covered the irrigation is interrupted to allow the animal to feed, and these periods last for about thirty minutes. *Arenicola* occupies about one-fifth of its tube and it has been estimated that the amount of oxygen dissolved in the sea water in the tube is about equal to that bound by the haemoglobin; the total amount at the end of an irrigation phase provides a reserve of about two hours, assuming that the oxygen tension of the venous blood can be reduced to a very low level. The high oxygen affinity will allow the pigment to continue operating until the oxygen tension in the sea water of the tube has fallen to a low level and so this animal would thus seem to be well adapted to conditions in which the oxygen tension in the environment varies over a wide range. The question as to whether or not this pigment acts as an oxygen store depends really on how the term is defined. It should probably be restricted to those pigments which operate under conditions in which a transport function ceases to be possible and in which the animal can only utilise the oxygen already bound by its pigment. However, since *Arenicola* periodically lives in a discrete, closed environment, the amount of oxygen contained in its blood and in the water of its tube may be thought of as a combined oxygen store.

The situation in the aquatic larva of *Chironomus plumosus* is almost identical to that in *Arenicola*. The animal lives in a tube which, like *Arenicola*, it irrigates periodically with pauses for filter-feeding or resting. During these periods the oxygen tension in the tube decreases, but because of its high affinity for oxygen (p_{50} is less than 1 mm) the pigment can continue to transport oxygen. If the larva goes into oxygen debt the transport function of the pigment helps this to be rapidly paid off when the animal starts irrigating again. This is particularly important if the water outside the tube has for some reason a low pO_2. In this animal there is additional evidence for a storage as well as a transport function in that, even at low oxygen tensions of the surrounding water, it takes several minutes for the oxyhaemoglobin to be completely deoxygenated. This dual function is particularly important in the maintenance of filter-feeding at low oxygen tensions, since this activity consumes a considerable amount of energy and requires aerobic conditions (table 5.3c).

In all of the above examples the haemoglobin is a high affinity pigment present in the blood plasma. It acts to transport under conditions of low environmental pO_2, the principal variable being the manner in which this low pO_2 occurs, whether it be a result of the environment, a high resistance at the respiratory surface, or periodic activity in a tube-dwelling animal. Also in some instances there is evidence that it acts partly as an oxygen store.

There are a few cases, however, in which haemoglobin acts as a low affinity respiratory pigment. Some burrowing polychaetes such as *Nephthys hombergii* and *Eupolymnia* fall into this category (figure 5.14). *Nephthys* differs from *Arenicola* in one very important respect: its burrows are temporary and collapse when the sand is exposed at low tide, leaving the animal surrounded only by interstitial water which at best has an oxygen tension of about 7 mm. Thus the oxygen store in the water of the burrow to which *Arenicola* has access is not available to *Nephthys*. Also the oxygen tension in the interstitial water is too low to be usable, since there is a gradient between the water and the arterial blood which is thought to be in the region of 10 mm (i.e. more than the tension in the interstitial water).

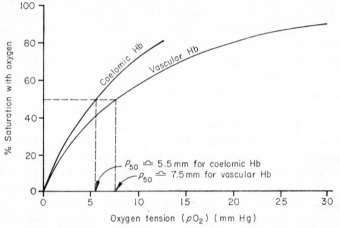

FIGURE 5·14 *Nephthys hombergii:* haemoglobin occurs in both the coelomic fluid and the blood. At the normal pH of 7.4 the affinity of the coelomic haemoglobin is higher than that of the vascular haemoglobin. However, the latter has a normal Bohr effect while the coelomic haemoglobin has a reversed one, and so at pH 7.0 there is no difference between the dissociation curves. The pO_2 of the interstitial water is about 7 mm. [After Jones (1955) *J. exp. Biol.*, **32**]

Thus a high affinity pigment would be useless. *Nephthys* is also interesting in that it possesses haemoglobin in the coelomic fluid as well as in the plasma. This behaves in the same way but the dissociation curve is displaced to the right (table 5.2c and figure 5·14).

The water bug *Anisops pellucens* (Notonectidae) presents another example of a haemoglobin with a low affinity for oxygen ($p_{50} = 28$ mm at 24°C, with no Bohr effect). This is of particular interest, not only because the haemoglobin is contained within cells, but because the cells themselves are in groups in the abdomen and each group is richly supplied with numerous tracheal branches (chapter 4). This animal periodically dives below the surface of the water, each dive lasting about four minutes at 25°C. A large part of each dive is spent in a condition of neutral buoyancy, a phenomenon typical of *Anisops* and *Buenoa* (both members of the Anisopinae) and which enables the animal to remain in mid-water with little or no expenditure of energy.

The thoracic and abdominal spiracles all open into a temporary air store (p. 95). This is in the nature of a physical gill, enclosed under long hairs on the ventral surface of the abdomen and in continuity with an air bubble trapped between the bases of the legs. The air bubble extends forwards and under the pronotum and wings. However, the evidence indicates that during a dive the oxygen bound by the pigment is of greater importance than that present in the air store. In the early part of the dive it is likely that the oxygen in the air store is utilised until its pO_2 falls to about 36 mm, at which point the oxygen bound by the pigment starts to be unloaded into the tracheae and some of it then passes into the air store. This is thought to be the onset of the phase of neutral buoyancy which is terminated when the pigment is maximally unloaded (represented by a pO_2 in the air store of about 20 mm). The animal now has to make active movements to maintain its level and soon returns to the surface.

The respiratory pigment is thus utilised entirely as an oxygen store and is not involved in oxygen transport. The physical gill, which has a less extensive area of contact with the water than that in some other Notonectidae (e.g. *Notonecta*), serves in part as an oxygen store, but mainly mediates the transfer of some of the oxygen released by the pigment to the thoracic spiracles (table 5.2d).

There is one other case in which the oxygen affinity is fairly low, and this is in members of the family Echiuroidea. These are annelid-like marine worms with a simple circulatory system devoid of a respiratory pigment; the haemoglobin is restricted to the coelomic fluid and is contained within corpuscles. In *Urechis* for example the value for p_{50} is 12.5 at 19°C and the oxygen capacity of its blood is 3.5 vol per cent. There is no Bohr effect, but the temperature effect is normal.

Urechis lives in muddy sand between the tide-marks in a U-shaped burrow, somewhat similar to that of *Arenicola*. Unlike *Arenicola* and *Chironomus* irrigation of the burrow and feeding are not mutually exclusive, but *Urechis* does have periods of twenty minutes or more of complete inactivity. In well-aerated water the pO_2 of the coelomic fluid is at least 75 mm and at this tension the pigment is fully saturated. However, the oxygen requirements of the animal are so small that the dissolved oxygen is more than adequate to meet its needs, and so while the animal is irrigating its burrow the pigment will be functionless. It has been calculated that the coelomic fluid contains sufficient oxygen for about one hour (less than ten minutes of this provided by the dissolved oxygen) without irrigation. Also, because of the low oxygen affinity of the pigment, most of this will be available before the pO_2 of the coelomic fluid falls below 10 mm. However, the water in the burrow contains sufficient oxygen for over two hours without irrigation and thus there may be very little unloading of the oxygen bound by the pigment during the short periods of rest. As far as the periods of exposure at low tide are concerned the oxygen tension in the burrow reaches a minimum of 0.96 vol per cent (equivalent to a pO_2 of 14 mm) during the fourth hour of exposure. After this it increases as the tide begins to rise again, but even at 14 mm the pigment is still 60 per cent saturated and so may be capable of transporting oxygen. Thus the oxygen requirements of *Urechis* are well served by a low affinity respiratory pigment. During irrigation of the burrow its oxygen-combining properties are not utilised, but during intervening periods it acts as an oxygen store and presumably also serves to transport oxygen from the water in the burrow (table 5.2c).

5.5. Haemerythrin The members of another small phylum of animals, the Sipunculoidea, also possess a respiratory pigment contained in corpuscles in the coelomic fluid. In this case it is haemerythrin, and it is also found in the tentacular canals. The coelomic pigment has a fairly low affinity for oxygen and is very similar to the haemoglobin of *Urechis*. In *Sipunculus* it has a somewhat lower p_{50} than *Urechis*: 8 mm at 19°C. It also has a lower oxygen capacity: only 1.6 vol per cent. Again there is no Bohr effect and the temperature effect is normal. The pO_2 of the coelomic fluid under well-aerated conditions is only about 20 mm, presumably because of the lack of a specialised respiratory surface, but even at this low tension the pigment is 85 per cent saturated at 19°C.

Sipunculus, like *Urechis*, is found in muddy sand, but it is rarely exposed at low water. The available evidence indicates that the coelomic pigment operates in a similar way to that in *Urechis*: that is, while the animal is in well-aerated conditions (in this case near the surface of the sand) the pigment remains non-functional, but is almost fully saturated with oxygen. When the animal is in more adverse conditions lower in the sand (it may penetrate to a depth of at least 30 cm) where the pO_2 is less, the pigment acts as an oxygen store and unloads its oxygen. Insufficient is known about *Sipunculus* to say whether or not the pigment can be used to transport oxygen from the surrounding water under these latter conditions (table 5.2c).

In some sipunculids the tentacular and coelomic respiratory pigments have different oxygen affinities. In *Dendrostomum* the tentacles are an important site of respiratory exchange and the coelomic pigment has the higher oxygen affinity, presumably so that oxygen can be transferred to it from the tentacular pigment (figure 5.15a). In the burrowing *Siphonosoma* the tentacles are not of particular respiratory importance, uptake of oxygen occurring across the body wall, and the situation is reversed, with the tentacular pigment having a slightly higher oxygen affinity (figure 5.15b).

5.6. Myoglobin Apart from its occurrence in vertebrate muscles (p. 103) myoglobin is found in the radular muscle of certain gastropod molluscs such as *Ischnochiton*, *Cryptochiton* and *Busycon*. It is of unit molecular size and each molecule combines with a single molecule of oxygen. The dissociation curve is almost hyperbolic and considerably to the left of that of the blood haemocyanin of these animals (figure 5.16). Thus the oxygen affinity is much higher than that of the haemocyanin and this greatly facilitates the unloading of oxygen from the blood to the myoglobin, so providing a particularly good oxygen supply to this very active muscle. There is virtually no Bohr effect.

5.7. Vanadium chromogen For the sake of completeness there is one other pigment which must be mentioned, chromogen. It occurs in a number of adult ascidians (sea squirts) such

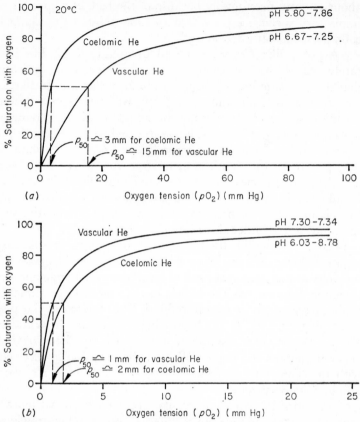

FIGURE 5·15 (a) *Dendrostomum zostericolum*. The coelomic haemery-thrin has a higher oxygen affinity than the vascular haemocyanin. Oxygen uptake is across the tentacles. (b) *Siphonosoma ingens*. The vascular haemerythrin has a higher oxygen affinity than the coelomic haemocyanin, although the difference is less than in *Dendrostomum*. Oxygen uptake is across the general body surface. The haemery-thrins of both *Dendrostomum* and *Siphonosoma* show virtually no Bohr effect. [After Manwell (1960) *Comp. Biochem. Physiol.*, **1**]

as *Ciona intestinalis* and is found associated with the so-called mulberry (green) cells of the blood. (In *Cynthia papillosa*, which does not have any mulberry cells, the pigment is absent.) There is no evidence for the presence of any bound oxygen in ascidian blood and the vanadium chromogen is thought to function as an

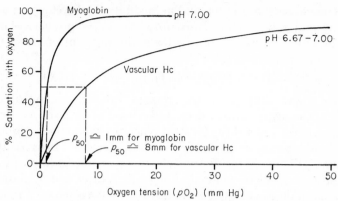

FIGURE 5·16 *Ischnochiton*. The oxygen affinity of the myoglobin of the radular muscle is much higher than that of the vascular haemocyanin. The latter shows a very small Bohr effect. [After Prosser and Brown (1961) *Comparative Animal Physiology*. Saunders: Philadelphia]

oxygen activator. However, little is known of its precise method of function.

5.8. The transport of carbon dioxide Relatively little is known at present about the transport of carbon dioxide in invertebrates. Apart from a certain amount passing into solution in the plasma it does combine reversibly with water to form carbonic acid

$$CO_2 + H_2O \rightleftharpoons H_2CO_3 \qquad (6)$$

In vertebrates this reaction occurs slowly in the plasma but the cells, in addition to containing the respiratory pigment (haemoglobin), contain carbonic anhydrase which increases the rate of formation of carbonic acid. It is a weak acid that tends to dissociate

$$H_2CO_3 \rightleftharpoons H^+ + HCO_3^- \qquad (7)$$

In the presence of one of its salts it will have the properties of a buffer, but this power will be small in most bloods (pH about 7.4 or more) since the acid dissociation constant is $10^{-6.15}$.

Haemoglobin includes a protein and is therefore amphoteric, that is it can act as an acid or a base depending on the hydrogen ion concentration (pH); the pH above which it will act as an acid is known as the isoionic point. The isoionic point of haemoglobin is at pH 6.8 and in blood it will therefore act as a weak acid and have buffering properties in the presence of one of its salts. In vertebrates, at least, the pigment exists in the form of a potassium salt. Thus

$$KHb + H^+ + HCO_3^- \rightleftharpoons HHb + K^+ + HCO_3^- \qquad (8)$$

Since the resultant acid is very weak little dissociation occurs and so the pH is hardly affected. Removal of the hydrogen ions increases the rates of the forward reactions in equations (6) and (7) and thus more carbon dioxide is absorbed. Because this occurs principally inside the cells there will be an increase in their content of bicarbonate ions with respect to the plasma. To maintain the balance this must be counteracted by a movement outwards of these ions (which would allow still more carbon dioxide to be absorbed) or by an inward movement of chloride ions

$$\frac{[HCO_3^-]_{cell}}{[HCO_3^-]_{plasma}} = \frac{[Cl^-]_{cell}}{[Cl^-]_{plasma}} \qquad (9)$$

Since the cell membrane is relatively impermeable to sodium and potassium ions, both of these movements occur to prevent a charge from forming on the cell. This exchange is called the 'chloride shift'. The plasma proteins (and phosphates) can act in a similar manner, but in invertebrates with haemoglobin or haemocyanin there is virtually no other protein present in the blood, and so only the respiratory pigments have any marked buffering effect.

Again in vertebrates a small amount of carbon dioxide combines with free amino groups of the protein molecules to form carbamates

$$CO_2 + Pr.NH_2 \rightleftharpoons Pr.NHCOOH \qquad (10)$$

Like oxygen, carbon dioxide has a dissociation curve, but the shape is generally hyperbolic (figure 5·17). The oxygen tension normally has an effect on the combination of carbon dioxide with haemoglobin. Since deoxygenated haemoglobin is a weaker acid than haemoglobin it has less tendency to dissociate electrolytically and release hydrogen ions, and thus acts as a more effective buffer. It therefore has a higher affinity for carbon dioxide. This means that in the tissues, where oxygen is given up, the absorption of carbon dioxide by the haemoglobin is facilitated and conversely at the respiratory surfaces, where the haemoglobin is oxygenated, carbon dioxide is driven off. This 'Haldane effect' is thus the counterpart of the Bohr effect. Haemocyanin behaves in a similar way and the haemocyanins of *Loligo* and *Limulus*, for example, have reasonable buffering capacities. The haemocyanin of *Loligo* shows a normal Haldane effect, but in animals which show a

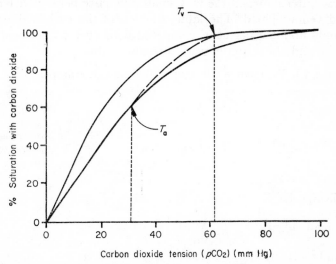

Carbon dioxide tension (pCO_2) (mm Hg)

FIGURE 5·17 *Carbon dioxide dissociation curve*. This is normally fairly hyperbolic. Increase in oxygen tension moves the curve to the right (Huxley effect) so that the true dissociation curve is represented by the dotted line between the CO_2 tension in the venous blood (T_v) and that in the arterial blood (T_a)

reversed Bohr effect within their physiological range, the Haldane effect may be absent (i.e. oxygenation has no effect on buffering) as in *Limulus*, or it may be reversed (i.e. oxygenated haemocyanin has a greater affinity for carbon dioxide) as in *Busycon*.

In air the partial pressure of carbon dioxide (pCO_2) is 0.23 mm mercury. Sea water in equilibrium with air would dissolve 0.2 vol per cent carbon dioxide but, because the carbon dioxide can combine with cations from various buffers (mainly as bicarbonate), it may in fact contain as much as 4.8 vol per cent.

In *Loligo* the carbon dioxide content of arterial blood is

3.98 vol per cent, equivalent to a pCO_2 of 2 mm mercury, and this increases to 8.27 vol per cent ($pCO_2 = 6$ mm) in the venous blood. The carbon dioxide gradient across the respiratory membranes is thus rather small. It is similarly small in decapod crustaceans; the corresponding figures for *Panulirus* for example are, at 15°C, arterial blood 10.00 vol per cent, venous blood 10.32 vol per cent. What is particularly striking about these latter figures is the very small difference between arterial and venous blood. The labile fraction of the carbon dioxide is only about 3 per cent, compared with over 50 per cent in *Loligo* and about 40 per cent in *Octopus* (table 5.4).

TABLE 5.4. Parameters affecting carbon dioxide transport

	Carbon dioxide content		Labile* fraction (%)	pH differences between arterial and venous blood
	Content (vol %)	pCO$_2$ (mm Hg)		
Loligo pealei	A3.98	2.2	52	
	V8.27	6.0		
Octopus vulgaris	A3.94		40	
	V6.56			
Panulirus interruptus	A10	5	3	0.02
	V10.32	5.4		
Loxorhynchus grandis	A20	18	1	0.01
	V20.2	18.5		
Homarus americanus	A6	2.3	3.5	0.03
	V6.22	2.5		
Busycon canalicalatum	A12.8	2.0	10.5	0.17
	V14.3	3.3		

A, Postbranchial; V, Prebranchial
* Calculated
Data from Henderson (1928) *Blood, a Study in General Physiology*. Univ. Press: Yale. (*Busycon*); Redfield and Goodkind (1929) *Brit. J. exp. Biol.*, **6** (*Loligo*); Redmond (1955) *J. cell. comp. Physiol.*, **46** (*Panulirus, Loxorynchus* and *Homarus*); and Wolvekamp in *Haemaglobin*. Butterworth: London (*Octopus*)

For some reason the blood of *Busycon* has a high carbon dioxide content for an invertebrate, (14.3 vol per cent in venous blood). The labile fraction of this is 10.3 per cent, which is similar to that of human blood, although the total carbon dioxide content of the latter is considerably higher (58 vol per cent in venous blood), presumably due to the presence of carbonic anhydrase in the cells.

It might be thought that the situation in an animal such as *Urechis*, with its haemoglobin contained within corpuscles, would approximate to that in mammals. However, the plasma of the coelomic fluid in this animal is practically protein-free and has no buffering power. Nevertheless it contains about half of the total carbon dioxide content at physiological tensions. There is evidence that carbonic anhydrase does occur within the cells, but that any chloride shift that may be present is limited.

Other evidence for the presence of carbonic anhydrase in invertebrate circulatory fluids is very sparse. It has been demonstrated in the coelomic fluid of *Sipunculus* and *Arenicola*, and in the blood of an earthworm and of the polychaete *Nereis*. On the other hand it does occur in many cases in the tissues, and particularly in the gills of aquatic animals. This means that the reactions of equations (6) and (7) occur in a predominantly forward direction in the tissues themselves and in the reverse direction in the gills. This contrasts with the situation in mammals, where the reverse reaction takes place in the blood stream, and resembles more what happens in the aquatic vertebrates.

Finally, in insects with an open tracheal system some at least of the carbon dioxide may be lost through the intersegmental membranes; although in general in this group most is probably lost through the spiracles.

SIX
Circulatory Systems

The circulatory system provides the important link between the environment and the cells. With increase in body size the coelom becomes incapable of supplying the circulatory needs of the animal and so the pathway of oxygen to the tissues must be facilitated. This has been achieved in many cases by the development and elaboration of a system of sinuses or vessels which contain a fluid (blood). A contractile mechanism is usually developed to pump the blood around the body. In some cases the larger vessels fulfil this function, but in others a special organ, the heart, has been developed. Pumping may be assisted by accessory booster hearts, especially in large or active animals, and also by body movements. Indeed, many small animals rely on body movements alone to circulate the blood and in these the circulatory system is poorly developed and specialised respiratory structures are absent. The function of the blood is to carry oxygen, usually in combination with a respiratory pigment, to the tissues and to remove the waste products of respiration.

The development of an efficient circulatory system is generally linked with the development of restricted sites of oxygen uptake. It is clear that an efficient circulatory system, especially one that is well-developed in the respiratory organ itself, will improve the efficiency of the respiratory organ by removing oxygen from the site of uptake as soon as it enters the animal, thus preserving the concentration gradient across the respiratory membrane.

Circulatory systems may be open or closed, although the dividing line between the two is somewhat tenuous. In an open system the circulatory fluid is only partly contained within discrete vessels and it passes into body spaces. These may be

136

large and fairly well defined (sinuses), or small and associated with the body wall or organs (lacunae). In a closed system the blood is entirely enclosed within discrete tubes which become very small in the region of the body wall and organs (capillaries).

Whereas in most animals the circulatory system is of prime importance for the transport of oxygen, this is by no means always the case. In some arthropods (e.g. insects) the oxygen is conveyed directly into the close proximity of the tissues and even of individual cells by the tracheae and tracheoles (chapter 4), but even so, a simple circulatory system is usually present. Furthermore some echinoderm groups are unusual in that, although a so-called haemal system is present, the coelomic fluid is almost certainly of greater importance in transporting oxygen.

6.1. Nemertini The nemertines do not possess a coelom and they are the only acoelomates to have a circulatory system. It is a closed system and consists of vessels, the largest of which are contractile, and mesenchymal spaces (lacunae). (This is a different use of the term lacunae, since these mesenchyme spaces are membrane-bounded.)

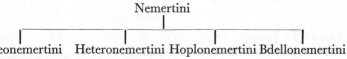

The circulatory system is at its simplest in the Cephalothricidae (Paleonemertini) in which there are two longitudinal lateral vessels joined anteriorly and posteriorly by lacunae (e.g. *Cephalothrix*). However, in most paleonemertines additional vessels are present, usually in the form of elaborations of the cephalic or anal lacunae and vessels to the proboscis cavity (rhyncocoel) (e.g. *Tubulanus*). In some paleonemertines and in the other three orders a longitudinal dorsal vessel occurs (e.g. *Amphiporus*) (figure 6·1; plate V) and, with the exception of the Hoplonemertini, there are lateral branches to the foregut (e.g. *Cerebratulus*). In the Bdellonemertini there are also branches to the rest of the digestive tract and to other parts of the body (e.g. *Malacobdella*) (figure 6·1). The branches to the foregut are of particular interest since some nemertines pump water in and out of the foregut, thus making it a region of respiratory significance. No valves occur in the vessels and, although the largest vessels are contractile, blood flow is not unidirectional.

6.2. Annelida The annelids have a true coelom, and in the polychaetes, the oligochaetes and one leech (*Acanthobdella*) this is very extensive and provides a hydroelastic skeleton for the animal. The blood system is closed and there are two main longitudinal vessels, a dorsal one in which the blood flows anteriorly and a ventral one in which it flows posteriorly. The dorsal vessel is thought to have arisen from a large sinus (the supposed primitive condition). Such a sinus is found in the oligochaete *Aelosoma*, except at the anterior end (figure 6·2).

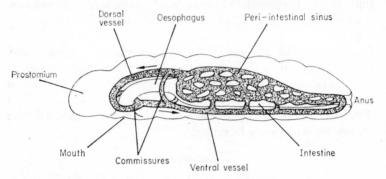

FIGURE 6·2 Circulatory system of the oligochaete *Aelosoma headleyi*. Arrows indicate direction of blood flow. [After Avel (1959) in *Traité de Zoologie*, Vol. V, (Ed. Grassé). Masson: Paris]

However, this is a small animal and it may be a secondary specialisation associated with size. In *Aelosoma* the dorsal and ventral vessels are linked anteriorly by two pairs of commissures, and more posteriorly by two or three other vessels.

However, in most polychaetes and oligochaetes there is a pair of lateral vessels arising from the ventral vessel in each segment. This lateral pair takes blood to the body wall, to the nephridia and (in polychaetes) to the parapodia, the latter being of particular respiratory significance in that they are often modified in part as gills (chapter 2). The blood is returned segmentally to the dorsal vessel. In addition a pair of intestinal vessels runs between the longitudinal vessels in each segment. In oligochaetes the blood in the intestinal trunks flows from ventral to dorsal, but it generally flows in the reverse direction in polychaetes. In many sedentary polychaetes (e.g. *Arenicola, Amphitrite, Pomatoceros*) the vessels around the anterior part of the gut and the dorsal vessel in this region are replaced by a large gut sinus (figure 6·3). It is not known

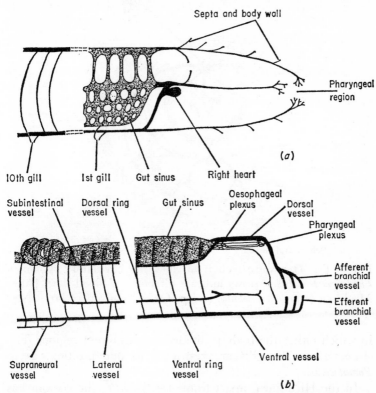

FIGURE 6·3 Circulatory systems of the polychaetes *Arenicola* (*a*) and *Amphitrite* (*b*). [*a* after Barnes (1968) *Invertebrate Zoology*. Saunders: Philadelphia (after Bullough); *b* after Thomas (1940) *L.M.B.C.* Memoirs, **33**]

whether this is a retention of the primitive condition or a secondary specialisation.

In many species the basic plan of the vessels just described is complicated by the presence of additional vessels. In *Lumbricus*, for example, a subneural and a pair of lateral neural vessels are present and in each segment these are joined by transverse branches (figure 6·4). Also the lateral neurals connect with the transverse vessels arising from the ventral vessel, and the subneural is connected to the dorsal longitudinal. The system becomes particularly complex in sedentary polychaetes

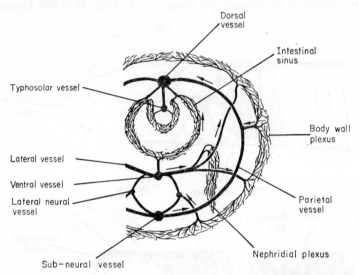

FIGURE 6·4 Transverse section to show the circulatory system of *Lumbricus terrestris*. Arrows indicate direction of blood flow. [After Meglitsch (1967) *Invertebrate Zoology*. Oxford University Press: New York]

in which either the body is divided into different regions (e.g. *Arenicola*) or the gills are restricted to one position, as in *Pomatoceros*.

In the Hirudinea, apart from *Acanthobdella*, the coelom has become reduced to a series of interconnecting sinuses. Only in the rhynchobdellids do blood vessels persist and here they only consist of a dorsal and a ventral vessel connected by looped

FIGURE 6·5 Hirudinea: Rhynchobdellida. (*a*) the arrangement of blood vessels in *Hemiclepsis marginata*; (*b*) the relationship between the reduced coelomic system and the dorsal blood vessel in *Piscicola geometra*; the contractile vesicles are also shown. The arrows denote the direction of blood flow. [After Harrant and Grasse (1959) in *Traité de Zoologie*, Vol. V, (1) (Ed. Grassé). Masson: Paris]

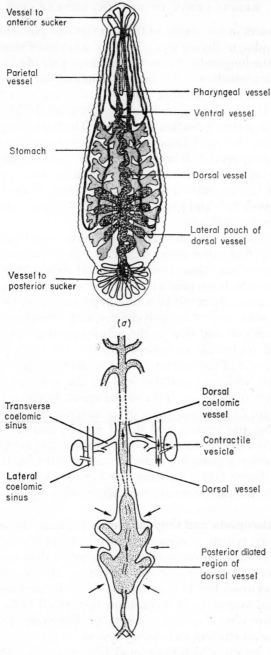

Vessel to anterior sucker

Parietal vessel

Pharyngeal vessel

Ventral vessel

Stomach

Dorsal vessel

Lateral pouch of dorsal vessel

Vessel to posterior sucker

(a)

Transverse coelomic sinus

Dorsal coelomic vessel

Contractile vesicle

Lateral coelomic sinus

Dorsal vessel

Posterior dilated region of dorsal vessel

(b)

commissures in the region of the anterior and posterior suckers and oesophagus (figure 6·5a). There is a coelomic sinus around each of the longitudinal vessels, and also a pair of lateral and a pair of intermediate (also lateral) longitudinal sinuses which connect with one another and with the dorsal and ventral sinuses. In the pharyngobdellids and gnathobdellids the coelomic sinus system is further developed. The lateral longitudinal sinuses are the most important, the dorsal and ventral ones having disappeared in some species. In rhynchobdellids the subepidermal plexus consists of numerous lacunae connecting with the lateral sinuses, and thus the system is open; but in pharyngobdellids and gnathobdellids the lacunae are replaced by coelomic capillaries.

In polychaetes and oligochaetes the dorsal vessel is the principal pump, but other major vessels are also contractile. Peristaltic waves pass along these vessels and these, combined with the valves which are present in the dorsal vessel in oligochaetes at least, normally result in a unidirectional blood flow. However, in some sedentary polychaetes with branchial tentacles there is an ebb and flow in the major vessels to the tentacles. Enlarged contractile chambers ('hearts') are often present in various parts of the system. Thus, in *Arenicola* there is a pair associated with the anterior end of the gut sinus, and in *Chaetopterus* there is one in the dorsal vessel. In oligochaetes the 'hearts' are formed from some of the anterior transverse vessels connecting the ventral and dorsal vessels, and they contain valves. Among the leeches the piscicolids (Rhynchobdellida) have a series of pumping chambers associated with the lateral sinuses (figure 6·5b). In those leeches without any vascular system the lateral sinuses are contractile and pump the coelomic fluid.

6.3. Arthropoda and Onychophora Typically the arthropod circulatory system is open and in all except the insects the organs are surrounded by channels (lacunae) derived from the haemocoel. Reduction or simplification of the circulatory system occurs either in conjunction with the development of a tracheal system or in association with small body size, the latter state obviating the necessity of developing respiratory organs or an efficient circulatory system.

In the insects, which have an elaborate tracheal system, the circulatory system is comparatively simple. The body cavity is a haemocoel divided by diaphragms into three main sinuses, which are all in continuity with one another. There is a dorsal pericardial sinus, a ventral sinus containing the nerve cord and a large perivisceral sinus in which lie the organs. The insect heart is extremely simple. It is modified from a dorsal vessel and lies in the abdominal part of the pericardial sinus. It is contractile and pumps blood forwards into an anterior aorta, from where it passes to the perivisceral and ventral sinuses. In some insects the flow of blood in the heart itself can be reversed. Blood enters the legs from the ventral sinus and returns to the perivisceral sinus. It moves dorsally in the abdomen into the pericardial sinus and enters the heart through paired openings called ostia, which are guarded by valves. In addition to this, booster hearts occur in some insects, for example at the bases of the wings or legs, and these aid the circulation of blood around the body.

The arrangement in the members of the Onychophora, which also have a tracheal system, is very similar to that in insects. The heart is very long and extends from the next-to-last segment of the body anteriorly as far as the first pair of legs. There is one pair of valvular ostia in each segment. The ventral sinus is replaced by two lateral sinuses, each containing a nerve cord and nephridia.

Many arthropods apart from insects have a long tubular heart, namely Merostomata, Diplopoda (millipedes), Chilopoda (centipedes), many arachnids and a number of crustaceans. In the millipedes and centipedes the heart extends almost the full length of the animal with two pairs of ostia per segment in the former and one pair per segment in the latter. They also possess lateral segmental arteries, which have their origin in the heart. A tracheal system is present throughout these two groups, with the exception of the scutigeromorph centipedes, which possess seven pairs of tracheal lungs. A large number of short tracheal tubes arise from these and project into the pericardial cavity to oxygenate the blood.

The arachnids are a very good group in which to study the levels of development of the circulatory system. Within the arachnids some orders rely solely on tracheae for respiration and even at this level a variation in tracheal complexity can be seen, reaching its culmination in the Solifugae. In other orders tracheae are absent and the respiratory organs are book-lungs (Scorpionida, Uropygi, Amblypygi, Schizomida, Palpigradida). Finally, in the Araneida (true spiders) tracheae, book-lungs, or a combination of both are used in different families.

At its best developed, as in the Scorpionida, Uropygi and Amblypygi, the arachnid circulatory system consists of a dorsal contractile heart (from which arises an anterior and a posterior aorta and a series of lateral arteries), haemocoelic lacunae around the body organs, and a number of sinuses or veins. The heart lies in the pericardial cavity and is generally restricted to the opisthosoma. In the Uropygi it extends over nine segments (two in the prosoma) and has nine pairs of ostia; in the Scorpionida and Amblypygi it extends over seven and six pairs of segments respectively, with a corresponding number of ostia. The anterior aorta is well developed and has a number of branches which supply the prosoma, but the posterior aorta is comparatively small and only supplies the terminal region of the opisthosoma. The viscera of the opisthosoma receive blood from the lateral arteries, of which there are nine pairs in the Scorpionida and six in the Amblypygi. Both aortae and the lateral arteries have a valve at their junctions with the heart to prevent backflow of blood.

After passing through the haemocoelic lacunae the blood finds its way into two large, longitudinal ventral sinuses which are enlarged in the region of each book-lung. The blood passes through the leaves of the book-lungs and is taken to the pericardial cavity in large branchiopericardial veins. In scorpions there are seven pairs of these draining the four pairs of book-lungs. The blood then re-enters the heart through the ostia during diastole.

The above three orders include animals which are several centimetres long and in which respiration is solely dependent on book-lungs. In the Schizomida and Palpigradida, which also have book-lungs but which comprise only small animals, the circulatory system is reduced. The heart for instance extends over fewer segments, with consequently fewer ostia, but even so there are still five pairs in the Schizomida and four pairs in the Palpigradida.

With the exception of the Solifugae, the orders which rely exclusively on a tracheal system contain comparatively small animals, and their circulatory system is reduced, both in heart size and in the virtual absence of arterial branching. The pseudoscorpions and phalangids for example have at most three and two pairs of ostia respectively, while most acarines have no heart at all, but simply a system of lacunae. Both anterior and posterior aortae are present when there is a heart, but there are no lateral arteries. However, the ventral abdominal sinuses may be well developed. In the Solifugae the tracheal system reaches its peak of development among the arachnids. The heart is unusually well developed and extends over eight segments (two in the prosoma as in the Uropygi) and has one pair of ostia per segment, but the rest of the circulatory system is very reduced, consisting mainly of haemocoelic lacunae.

A relationship between type of respiratory organ and the number of heart ostia is particularly apparent in the true spiders (Araneida). In those spiders which have only book-lungs the heart has between three and five pairs of ostia, in those which only have tracheae there are two pairs, and in those which have a combination of pulmonary and tracheal respiration there are either two or three pairs. There are usually two or three pairs of lateral arteries present and, where book-lungs occur, the pulmonary veins are very well developed (figure 6·6). In spiders some of the blood is thought to return to the pericardial cavity without passing through the lungs.

It is thus apparent that the development of an efficient tracheal system, which carries the oxygen into the close proximity of the tissues, largely obviates the need for an efficient well-developed circulatory system. A further interesting point is that in some arachnids most of the major nerves run within arteries, and this is carried a stage further in the Merostomata, in which the nerve cord is similarly enclosed. The significance of this is not readily apparent.

The Merostomata (e.g. *Limulus*) possess respiratory organs called book-gills. They have a long heart with eight pairs of ostia guarded by valves, and their circulatory system is similar to, but rather more complex than, that of arachnids. Thus there are three anterior arteries arising directly from the heart, and the four pairs of lateral arteries enter a pair of collateral arteries which unite behind the heart. The anterior and collateral arteries give rise to a number of branches. Blood from the haemocoelic lacunae is collected in dorsal and lateral vessels and conducted to a pair of ventral collecting vessels and from these an afferent branchial vein passes to each gill. The aerated blood returns to the pericardial cavity, and hence the heart, via efferent branchial vessels and a branchiopericardial vein. The book-gills are moved backwards and forwards and

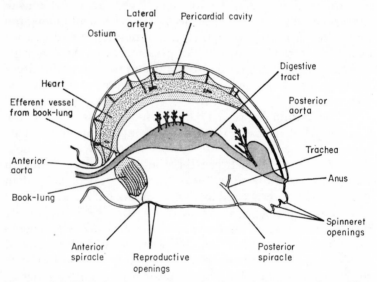

FIGURE 6·6 The circulatory system in the opisthosoma of a spider. [After Meglitsch (1967) *Invertebrate Zoology*. Oxford University Press: New York]

this both produces a respiratory current and acts as an auxillary pumping mechanism. When the gills move forwards blood is pumped into the lamellae; when they move backwards it is expelled.

Among crustaceans numerous smaller ones such as many of the ostracods and copepods do not have a heart; others such as cladocerans have a small globular heart with only a single pair of ostia and a short unbranched anterior aorta. Larger animals tend to have longer hearts with both anterior and posterior aortae but, apart from the Malacostraca, the remainder of the circulatory system is relatively simple. The heart is usually confined to the thorax, but in the Anostraca (fairy shrimps) it runs almost the full length of the body and in the Isopoda it is confined to the abdomen (presumably connected with the abdominal position of the gills in this latter order). Within the Malacostraca, the eumalacostracans, syncarids and hoplocarids have fairly simple circulatory systems with a long, tubular heart and a number of pairs of ostia. In the higher groups the heart tends to be shorter with a corresponding decrease in the number of ostia. Thus isopods have one to four pairs, amphipods one to three pairs and decapods three pairs. In the isopods

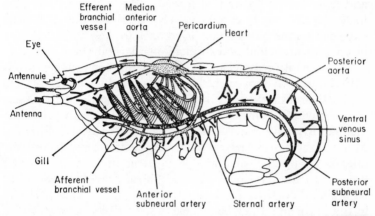

FIGURE 6·7 The circulatory system in the lobster. The arrows indicate the direction of blood flow. [After Gegenbauer (1878) *Elements of Comparative Anatomy* (Trans. Bell and Lankester). London: Macmillan]

and decapods in particular there is a highly developed arterial system and the circulatory system as a whole reaches a similar level of complexity to that found in the Merostomata and some arachnids.

In decapods several arteries leave the heart anteriorly and there is also a posterior aorta and a sternal (descending) artery (figure 6·7). The latter arises either directly from the heart or from the base of the posterior aorta, and divides below the nerve cord into anterior (thoracic) and posterior (abdominal) subneural arteries. Many crustaceans also have lateral arteries arising directly from the heart. The anterior arteries supply the head and thorax, and the posterior aorta the abdomen, while the main supply to the appendages is via the subneural arteries. The appendages are often divided by septa and so the blood is channelled within them; in decapods the arrangement is such that the blood vessels actually run along them.

Valves occur in various parts of the system, for example at the bases of the arteries and guarding the ostia. From the fine arterial branches the blood generally runs into a system of haemocoelic lacunae, but there is some development of capillary networks in decapods, notably in connection with the cerebral ganglion. The blood is ultimately collected in ventral sinuses and from here it passes into a median ventral sinus and then in to the afferent branchial vessels leading to the gills. After oxygenation the blood is conveyed back to the pericardial sinus in efferent branchial and branchiopericardial vessels.

Many small crustaceans which do not have a heart rely entirely on body movements to circulate the haemolymph, but in the Thoracica the rostral blood sinus is contractile. In many of those with a contractile heart the circulation may be augmented by body movements or by booster hearts. In most Malacostraca an enlargement of the anterior artery (cor frontale) is contractile, and analogous structures occur in other groups. In the terrestrial decapod *Coenobita*, vessels associated with the abdominal respiratory lacunae are contractile; and in the branchipod *Leptodora* there is a pair of structures called appendage organs, each consisting of a pulsating membranous disc between the first and second segments of the first thoracic legs.

6.4. Mollusca The circulatory system is generally open, as in the arthropods, and consists partly of blood vessels and partly of haemocoelic lacunae. True capillary networks are found only in some organs (e.g. the gills) of one class, the cephalopods. At the other extreme the Aplacophora and Scaphopoda (tusk shells) have a reduced circulatory system composed largely or entirely of a number of sinuses. The dorsal vessel, except in these latter two classes, is modified into a contractile ventricle enclosed in a pericardial cavity, but the heart differs from that encountered in the previous groups in that one or more pairs of auricles are also present. With the exception of the cephalopods these are fairly distinct from the ventricle. They are rather narrow tubular structures in the Monoplacophora but are better developed in the other groups.

The blood is typically pumped from the ventricle into an anterior aorta, from which it flows into arterial branches and is conducted to the haemocoelic lacunae surrounding the organs. It then runs via one or two large sinuses to the afferent branchial vessels which take it to the gills. It leaves the gills in the efferent branchial vessels which terminate in the auricles, and blood re-enters the heart during diastole.

In both the Polyplacophora and the Monoplacophora the

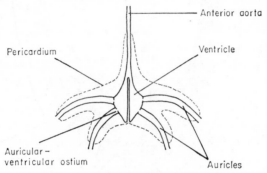

FIGURE 6·8 Heart of the monoplacophoran *Neopilina galathea*. [After Lemche and Wingstrand (1959) *Galathea Rep.*, **3**]

heart is posterior in position. The Monoplacophora (i.e. *Neopilina*) are unusual in that there is a pair of ventricles, each of which gives rise to a very short anterior aorta. These join in the mid-line to form a single longitudinal aorta (figure 6·8). There are two pairs of auricles, the anterior of which receives blood from the first four pairs of gills, the posterior from the last one or two pairs of gills.

In the Polyplacophora (chitons) there are a number of gills on each side. There is a pair of longitudinal gills veins and each one receives the blood from all of the efferent branchial veins on its side; there are only two auricles, one at the termination of each gill vein (figure 6·9).

In the other classes—Gastropoda, Bivalvia, Cephalopoda—there is one auricle at the base of each true ctenidium (i.e. not of secondary gills).

In gastropods the heart has become anteriorly situated as a

result of torsion. Some archeogastropod prosobranchs have two ctenidia and consequently two auricles (figure 6·10*a*), but in others the right ctenidium is reduced or absent and the right auricle in these is usually reduced (figure 6·10*b*). The mesogastropods and neogastropods all have a single (left) auricle

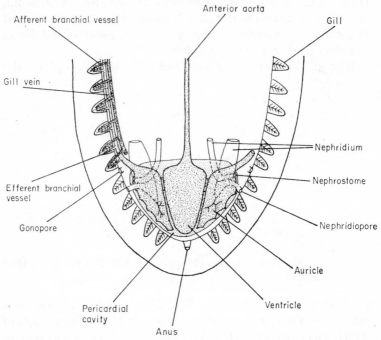

FIGURE 6·9 Posterior half of a chiton (Polyplacophora). The afferent and efferent branchial vessels are only drawn to six gills on the left side. [After Meglitsch (1967) *Invertebrate Zoology*. Oxford University Press: New York]

associated with the single ctenidium (figure 6·10*c*), as do some opisthobranchs (figure 6·10*d*). In other opisthobranchs and in pulmonates both ctenidia are lost but the left auricle is still retained.

In prosobranchs the auricle(s) lies anterior to the ventricle, but detorsion results in it assuming a posterior position again in opisthobranchs and pulmonates. Most gastropods have a posterior as well as an anterior aorta, and the latter divides into two main arteries, the cephalic and the visceral. The blood usually passes into lacunae around the kidney before reaching the gills or lung, but some may go direct to the gills and in certain pulmonates some may pass direct to the auricle from the kidney. There is a much more precise relationship between the circulatory and excretory systems than in the chitons for

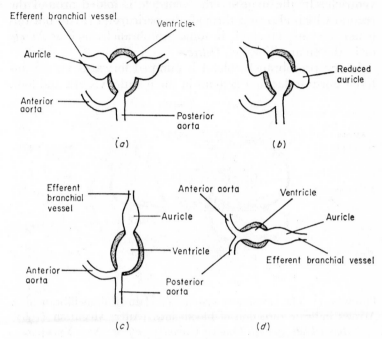

FIGURE 6·10 Hearts of various gastropods. (*a*) An archeogastropod with two gills. Note the posterior emergence of the aorta which is a result of torsion; (*b*) an archeogastropod with one gill and concurrent reduction of one auricle; (*c*) a mesogastropod with one gill and one auricle; (*d*) an opisthobranch showing the effect of partial detorsion. [After Meglitsch (1967) *Invertebrate Zoology*. Oxford University Press: New York]

example, in which the two nephridia simply ramify in the haemocoel and connect with the perivisceral cavity.

The blood in the ctenidia flows in the opposite direction to the respiratory current and thus makes full use of the concentration differences of the respiratory gases in the blood and in the respiratory medium to enable the maximum amount of

F

oxygen to be extracted from the water (countercurrent principle, chapter 1).

Bivalves typically possess two gills (with the exception of the septibranchs, which have two vascularised mantle cavities) and there are two auricles attached laterally to a single median ventricle. In the majority the ventricle is folded around the rectum which also runs through the pericardial cavity but this is not the case, however, in some protobranchs such as *Nucula* or in the eulamellibranch *Ostrea.*

In the protobranchs blood is pumped forwards via an anterior aorta to lacunar systems in the mantle, viscera and foot.

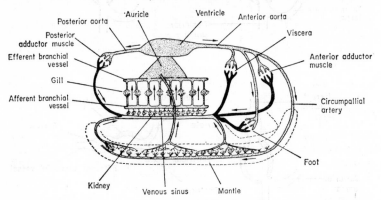

FIGURE 6·11 The circulatory system of a clam (Eulamellibranchia). Arrows indicate direction of blood flow. [After Meglitsch (1967) *Invertebrate Zoology*. Oxford University Press: New York]

It is then collected in sinuses and passes through the nephridia into the afferent branchial vessels. After oxygenation in the gills the blood runs into efferent branchial vessels and back to the auricles. In some (e.g. *Nucula*) there is a small posterior aorta in addition to the anterior one.

There is just an anterior aorta in some filibranchs (e.g. *Mytilus*), but most of this group have a posterior one as well, and in the eulamellibranchs the posterior is as well developed as the anterior. In many filibranchs and eulamellibranchs the surface area of the mantle is large and quite well vascularised, and consequently is important in respiration. This process of vascularisation of the mantle reaches its peak in the septibranchs, in which the gills have been lost. There are a number of variations on the basic protobranch-type circulation: thus part of the blood from the mantle may return direct to the auricle, while the remainder follows the usual pathway via the kidneys and gills (e.g. *Anodonta* and *Pecten*), or some of the blood may pass direct from the nephridia to the heart without passing through the gills (e.g. *Anodonta* and *Mytilus*) (figure 6·11).

Cephalopods possess either two pairs of gills, as in *Nautilus* (Nautiloidea), or only a single pair, as in the Coleoidea (octopods and squids). Consequently there are two pairs and a single pair of auricles respectively. The circulatory system is much more extensive than in the other molluscan classes and the haemocoelic lacunae are generally replaced by true capillary beds, thus giving a virtually closed system. These changes are presumably commensurate with the much more active life of these animals. Blood from the single ventricle is pumped along anterior and posterior aortae and distributed to the various organs of the body (figure 6·12a). It is ultimately collected in a series of veins which lead via anterior and lateral vena cavae into the afferent branchial vessels (figure 6·12b). The nephridia are in close association with branches of the anterior vena cava. After oxygenation in the gills the blood runs into the efferent branchial veins and thence to the auricles of the heart.

A booster heart occurs in some of the larger gastropods. It may be in the anterior aorta (aortic bulb), either in the pericardial cavity (e.g. *Patella*) or further forwards in the head region (e.g. *Busycon*). Cephalopods also possess booster hearts where they are typically situated at the base of each afferent branchial vessel (branchial hearts); in addition to this some of the arteries themselves are contractile.

6.5. Echinodermata Except in the crinoids and ophiuroids the coelom is well developed in this phylum and is of paramount importance in the circulation of the respiratory gases. However, there is a haemal (circulatory) system present which is composed of vessels lying in extensions of the perivisceral coelom called perihaemal sinuses (figure 6·13). In the asteroids the haemal system consists of an oral and an aboral ring, both with radial branches (or lacunae), connected by a large structure known as the axial gland which contains haemal pathways. Associated with this gland is a special part of the coelom, the dorsal sac, which is contractile. There are sinuses in the wall of the cardiac stomach and these communicate with the haemal plexus of the axial gland via the gastric haemal

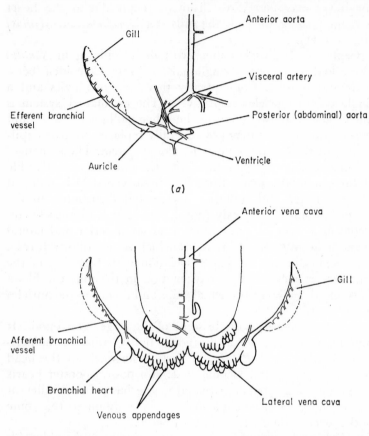

FIGURE 6·12 Main arteries (*a*) and veins (*b*) of the cephalopod *Eledone*. [After Isgrove (1909) *L.M.B.C. Memoirs*, **18**]

tufts, of which there are usually two. In these animals the papulae and podia are the main sites of respiratory exchange. The cilia of the papulae produce respiratory currents and, in addition, cilia on their internal surface circulate the coelomic fluid. Furthermore the coelomic fluid is also circulated by ciliary action in other parts of the coelom and is moved outwards along the aboral walls and inwards along the lateral walls.

The haemal system in ophiuroids is similar to that in asteroids, but is presumably more important in the conduction of the respiratory gases in the former because of the reduced coelom.

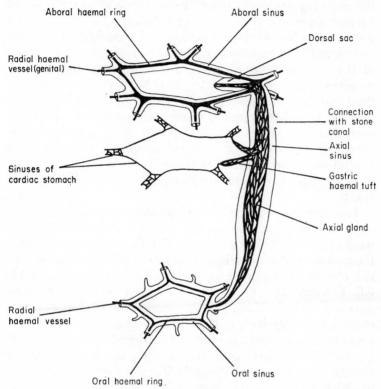

FIGURE 6·13 Haemal system of *Asterias*. [After Cuénot (1948) in *Traité de Zoologie*, Vol. XI, (Ed. Grassé). Masson: Paris]

Indeed there are branches to the bursal sacs which are used as respiratory organs.

In echinoids the axial gland lies free in the perivisceral coelom. There are two main haemal vessels arising from the oral ring, the inner and outer marginal sinuses, and in some cases there is a contractile collateral haemal sinus running parallel to the marginal sinuses for much of their length and connecting with the outer one by numerous branches. The oral radial lacunae give off branches to the podia, some of which are modified for respiration.

In the Holothuroidea there is no aboral ring and the axial gland is reduced or absent. As in echinoids podial branches arise from the radial vessels of the oral ring. Two vessels from the oral ring are associated with the gut: a contractile dorsal haemal sinus and a ventral haemal sinus (figure 3·5). The dorsal sinus gives rise to an interconnecting network of channels (rete mirabile) which is closely associated with the left branch of the respiratory tree. It has been suggested that oxygen passes from the terminal vesicles of this branch into the coelomic fluid and then into the haemal network. The blood from this network is collected into a vessel which parallels the dorsal sinus and then enters a haemal plexus in the intestinal wall. After passing through the intestinal sinuses it is collected in the ventral sinus and returned to the oral ring. Some blood does, however, pass directly to the intestinal plexus from the dorsal sinus.

The coelom is reduced to a series of interconnecting spaces in the crinoids, and the haemal system is composed of a network of sinuses within the connective tissue that has invaded the coelom. These animals, as may be expected from the reduction of both systems, have no specialised areas of oxygen uptake.

6.6. Echiuroidea With the exception of *Urechis* these animals have a closed circulatory system reminiscent of that of oligochaetes and polychaetes. Blood passes anteriorly in a dorsal vessel to the proboscis. It returns posteriorly in a ventral vessel lying above the nerve cord and re-enters the dorsal vessel via circumoesophageal and circumintestinal vessels, one or more of which are dilated and contractile.

6.7. Chordata The circulatory system in tunicates (Urochordata, Asidiacea) is entirely open. The heart is a short tube lying in a pericardial cavity and is formed by folding of the pericardial wall. The dorsal end of the heart is continuous with an abdominal sinus, and the ventral end with the subendostylar sinus. Branches from the subendostylar sinus connect across the pharyngeal region with the median dorsal sinus. The pharyngeal region itself is perforated so that water can pass through it, and it serves a respiratory function. The abdominal and median dorsal sinuses both terminate in the lacunar systems of the body. Blood flow can be in either direction, since peristaltic waves pass over the heart in one direction and then eventually slow down, stop and reverse direction.

6.8. Hemichordata Enteropneusts such as *Balanoglossus* have a system of sinuses and lacunae (figure 6·14). There is a central sinus which lies at the base of the proboscis and overlying this is a contractile body, the heart vesicle. Blood is pumped from the central sinus into a glomerular sinus and from there passes either anteriorly to the proboscis or via peribuccal vessels to a

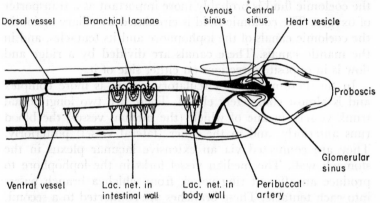

FIGURE 6·14 Circulatory system of an enteropneust (hemichordate) *lac. net.* lacunar network. [After Meglitsch (1967) *Invertebrate Zoology*. Oxford University Press: New York]

ventral vessel, which conducts it posteriorly. Blood from the proboscis returns to the central sinus by way of the venous sinus, which lies immediately behind the former. The blood in the ventral vessel passes into lacunar networks in the body wall and digestive system, and the network in the pharyngeal region supplies the branchial lacunae. As in tunicates the pharyngeal region is perforated by slits and has a respiratory function. The blood eventually enters a dorsal vessel and is carried anteriorly to the venous sinus. In addition to the heart vesicle both dorsal and ventral vessels are contractile. The system in the pterobranch (Hemichordata) *Cephalodiscus* is essentially similar, but in this animal even the main channels are thought to be lacunar, and this simplification may be related to small body size.

6.9. Other Phyla Two of the lophophorate phyla have circulatory systems. (The lophophore is a ridge around the mouth bearing tentacles with a coelomic lumen.) There are no organs exclusively specialised for respiration but the tentacles presumably are important in this respect in that they increase considerably the surface area of the animals.

The Brachiopoda are comparatively small animals enclosed in a bilaterally symmetrical bivalve shell. They have an open circulatory system which includes a dorsal contractile heart vesicle and blood vessels. Although there are tentacular vessels the coelomic fluid is probably more important as a transporter of oxygen. The coelomic fluid is circulated by ciliary action in the coelomic canals of the lophophore and its tentacles, and in the mantle canals. These canals are divided by a ridge and flow is in opposite directions on either side of it.

In the Phoronida the circulatory system is more complex and is almost completely closed. There are two longitudinal trunk vessels, in one of which (the median vessel) the blood runs anteriorly, and in the other (lateral vessel) posteriorly. They are connected via an extensive lacunar plexus in the stomach wall. The median vessel forks in the lophophore to produce an afferent ring vessel, from which a branch passes into each tentacle. These branches are connected to a second, efferent, ring vessel and from this two branches arise, run posteriorly and eventually join to form the lateral vessel. Blood flow is anterior in the median vessel and posterior in the lateral vessel, and short branches of the latter are highly contractile. Flow in the tentacular vessels is first in one direction and then in the other.

There remains only the phylum Pogonophora, a group of animals with tentacles and a well-developed, closed circulatory system including a dorsal and a ventral longitudinal vessel. Present evidence indicates a reversal of the usual direction of flow, with blood conducted anteriorly in the ventral vessel and posteriorly in the dorsal one, although there is some difficulty in deciding on the correct orientation of the body in these animals. Afferent tentacular branches arise from the ventral vessel and efferent tentacular vessels join the dorsal vessel, forming a single loop in each tentacle. The ventral vessel is enlarged in the protostome to form a muscular heart.

6.10. Control of the Heart Heart muscles may contract and relax rhythmically due to their inherently spontaneous contractile nature, and this is sometimes phased by a specific part of the heart wall which is dominant and acts as a pacemaker. Although contractions are not initiated by neurones the various parameters of the heart beat, such as frequency, amplitude and regularity, may all be modified by nervous activity. Such hearts are called myogenic, and are typical of molluscs. There is little good evidence of any arthropod having a myogenic heart, although their existence cannot be completely ruled out and there is some indication that they occur in scorpions, lower crustaceans and some insects (e.g. Diptera). The tunicate heart appears to be myogenic with a pacemaker region at each end of the heart; these pacemakers become dominant alternately. Control from the central nervous system may be only excitatory in nature, as is thought to be the case in prosobranch gastropods, but is generally both excitatory and inhibitory (e.g. pulmonate gastropods, bivalves and cephalopods).

Arthropods usually, if not typically, have a regulated, neurogenic heart, that is one in which beating is both initiated and controlled by nervous activity. Neurogenic hearts have been found in crustaceans, insects, spiders and *Limulus* (Merostomata) and possibly also in some gastropods (e.g. the prosobranch *Busycon*). Initiation of contraction is the role of ganglion cells which may be segmentally arranged, or grouped together into a definitive cardiac ganglion as in *Limulus* and most crustaceans. Furthermore, they may be divided functionally into pacemaker cells and follower cells. Regulation, both excitatory and inhibitory, takes place via extrinsic motor neurones from the central nervous system, and these may be segmental or reduced in number (three in decapod crustaceans). Most research has been carried out on decapod crustaceans, in which both initiation and regulatory mechanisms have reached a high level of development.

The cephalopod heart has a ganglion on each auricle, but these are concerned with regulation rather than initiation, and the heart is myogenic. Stretch controls the system in an excitatory capacity, and a certain degree of stretch is required to coordinate the auricles and ventricle.

Nervous control generally also extends to the valves, alary

muscles and, in cephalopods, to the accessory branchial hearts. The alary muscles suspend the heart from the roof of the pericardial cavity and contraction of them helps diastole. Apart from nervous control hearts may also be subject to hormonal control, as for example in the Crustacea, where the pericardial organs assume this function.

SEVEN
Intracellular Respiration

The previous chapters have been concerned with those structures involved in the extraction of oxygen from the environment and the means by which this oxygen is transported to the tissues to enable energy to be produced from the breakdown of nutrients ingested by the animal. While it is beyond the scope of this book to give a detailed consideration of the biochemical processes involved it is necessary for completeness to consider their nature in outline. Indeed they are known in detail only in the vertebrates and it is from these rather than from the invertebrates that much of the basic information is drawn.

Carbohydrates, proteins and lipids are the basic substrates that can be broken down to produce energy. The breakdown is an oxidation process in so far as atmospheric oxygen is required for its completion, although the initial stages can proceed in the absence of oxygen (i.e. anaerobically).

7.1. Carbohydrates Carbohydrates occur as monosaccharides (e.g. glucose), disaccharides (e.g. sucrose and trehalose), trisaccharides and polysaccharides (e.g. glycogen and starch). The di-, tri- and polysaccharides are initially broken down into less complex molecules, glycogen for instance being broken down and phosphorylated to form glucose-1-phosphate. This phosphorylation does not require energy, only the presence of inorganic phosphate. Following this the glucose-1-phosphate molecule is rearranged to form glucose-6-phosphate. Glucose on the other hand is converted directly into glucose-6-phosphate, but this phosphorylation requires energy in the form of ATP. The fate of the glucose-6-phosphate molecule is as follows: first its structure is altered to form fructose-6-phosphate and this is further phosphorylated, with the aid of ATP, to fructose-1,6-diphosphate. Each molecule of fructose-1,6-diphosphate is then split into one molecule of glyceraldehyde-3-phosphate and one of dihydroxyacetone phosphate. These are both triose phosphates and the latter is readily rearranged to form glyceraldehyde-3-phosphate, so that each molecule of glucose can be looked upon as having been broken down into two molecules of glyceraldehyde-3-phosphate.

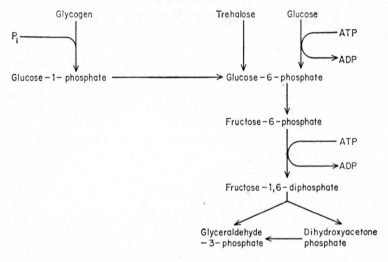

Carbohydrates are stored in the form of polysaccharides and, in both invertebrates and vertebrates, glycogen is the most usual storage form. In insects at least, carbohydrates are normally transferred to where they are needed, not as a monosaccharide such as glucose, but as the disaccharide trehalose. This substance is readily converted to glucose-6-phosphate and its high concentration in the haemolymph is presumably necessary for the high energy requirements of flying insects (see p. 91). However, trehalose is certainly not the transported form in all invertebrates, and in some decapod crustaceans another disaccharide, maltose, is probably important in this respect.

The glyceraldehyde-3-phosphate molecule undergoes a series of reactions to form pyruvic acid. This includes yet another phosphorylation (using inorganic phosphate) and involves the formation of two molecules of ATP from ADP and the release of two hydrogen atoms, which are ultimately taken up by the electron transport chain. The sequence of events leading to the production of pyruvic acid is known as glycolysis and can proceed in the absence of oxygen. The pyruvic acid is converted to acetyl coenzyme A, involving a decarboxylation and the reduction of one molecule of lipoic acid for each molecule of pyruvic acid. The lipoic acid molecule is re-oxidised by passing the two hydrogen atoms involved to a hydrogen carrier and hence to the electron transport chain.

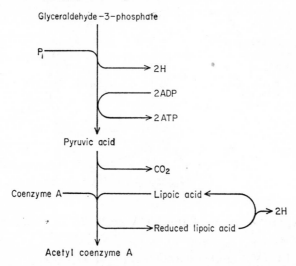

Thus one molecule of glucose is converted to two molecules of acetyl coenzyme A. In the process there is a net gain of two molecules of ATP and the liberation of eight hydrogen atoms to the electron transport system. The overall reaction so far can be summarised

Glucose $+$ 2 Coenzyme A $+$ 2 ATP$\rightarrow$2 Acetyl co-enzyme A $+$ 4 ATP $+$ 8 H $+$ 2 CO_2

Each molecule of acetyl coenzyme A then combines with one of oxaloacetic acid to form citric acid, which undergoes a series of reactions, eventually forming oxaloacetic acid again. This series is known as the tricarboxylic acid (TCA), Krebs or Citric acid cycle. At four separate points in this cyclical series of reactions a pair of hydrogen atoms is liberated to hydrogen-carrying enzymes and the electron transport system. Thus there are 16 atoms of hydrogen liberated here for each molecule of glucose. (In addition two molecules of carbon dioxide are liberated for each molecule of acetyl-coenzyme A that enters the cycle.)

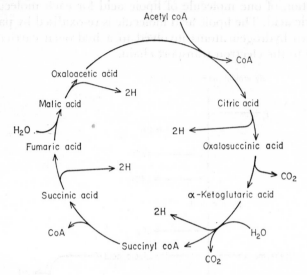

The overall reaction of the TCA cycle for one molecule of glucose is:

2 Acetyl CoA + 2 Oxaloacetic acid + 4 $H_2O \rightarrow$
2 CoA + 2 Oxaloacetic acid + 16 H + 4 CO_2

Altogether, for one molecule of glucose, two pairs of hydrogen atoms are released during glycolysis, two pairs in the formation of acetyl coenzyme A from pyruvic acid and eight pairs in the course of the TCA cycle; this gives 12 pairs in all. The two pairs produced in the formation of fumaric from succinic acid (in the TCA cycle) are passed either to FMN or FAD (hydrogen-carrying enzymes). The other ten pairs are passed to another hydrogen-carrying enzyme, in this case NAD* (Nicotinamide adenine dinucleotide) or NADP† (Nicotinamide adenine dinucleotide phosphate). These latter enzymes immediately relinquish the hydrogen to FMN or FAD.

FMN and FAD do not hand on the hydrogen atoms directly, but split them into hydrogen ions and electrons. The electrons

* or DPN (diphosphopyridine nucleotide).
† or TPN (triphosphopyridine nucleotide).

are passed to the electron transport chain. Ubiquinone receives the electrons and passes them to the cytochromes, of which there are several. The final step in this chain involves the enzyme cytochrome oxidase. This utilises each pair of electrons to activate an atom of oxygen to unite with a pair of hydrogen ions, forming a molecule of water. Although the precise details of this hydrogen transport system have not been worked out it is thought that, for each pair of hydrogen atoms entering the chain, three molecules of ATP are produced. Thus 36 are formed by this process from each molecule of glucose and so, with the net production of two molecules of ATP in glycolysis, this gives 38 molecules of ATP per molecule of glucose, representing about 342,000 calories of the 685,000 calories theoretically available. In other words cellular respiration is about 50 per cent efficient when the respiratory substrate is glucose.

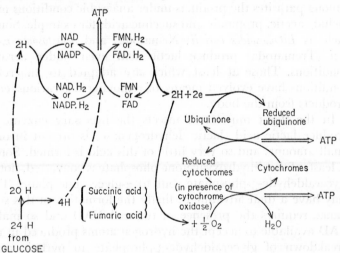

The general equation for the breakdown of a monosaccharide such as glucose is

$$C_6H_{12}O_6 + 6O_2 \rightarrow 6\ CO_2 + 6\ H_2O + 38\ ATP\ (342,000\ cals)$$
$$+ heat\ (343,000\ cals)$$

From this equation it can be seen that equal volumes of oxygen and carbon dioxide are involved. Their ratio (CO_2/O_2), which in this case is equal to one, is known as the Respiratory Quotient (RQ). Since for fats the RQ is 0.70 to 0.71 and for proteins it is 0.80 to 0.82, it is possible to make some deductions about the respiratory substrate of an animal from its respiratory quotient.

When an animal is particularly active, so that insufficient oxygen is available for all of the pyruvic acid to enter the TCA cycle, then it must go into a state of oxygen debt. In vertebrates this involves the formation of lactic from pyruvic acid with a consequent release of energy. This can also occur for example in crustacean muscle and in a number of insects which are prone to conditions of limited oxygen. The oxygen debt is 'paid off' when the animal returns to rest by an increase in ventilation, and any lactic acid that has accumulated is oxidised to pyruvic acid.

Although glycolysis is thought to occur in all animals there are a number in which the fate of the pyruvic acid differs. Among parasites the products under anaerobic conditions may include acetic, propionic and succinic acids for example. Some, such as *Litomosoides carinii* (Nematoda) and *Schistosoma mansoni* (Trematoda) produce lactic acid even under aerobic conditions. Those at least which are adapted to anaerobic conditions have evolved means of removing the various end-products from the body.

In the flight muscles of insects the necessary enzyme to produce lactic acid (lactic dehydrogenase) is present in only small amounts and so very little of this acid is formed. Some, at least, of the dihydroxyacetone phosphate is converted, not to glyceraldehyde-3-phosphate, but to α-glycerophosphate. This may have a dual advantage: firstly the formation of this substance requires the presence of reduced NAD and so makes NAD available to accept the hydrogen atoms produced in the breakdown of glyceraldehyde-3-phosphate to pyruvic acid. Secondly the α-glycerophosphate may be broken down without entering the TCA cycle, and its energy directly passed via hydrogen-carrying enzymes to the electron transport system. Under conditions of oxygen stress approximately equal amounts of pyruvic acid and α-glycerophosphate are formed. However, this does not increase the production of energy under these conditions in the way that lactic acid formation does. When the animal returns to rest the α-glycerophosphate may be reconverted to dihydroxyacetone phosphate and hence to pyruvic acid by the usual pathway. The hydrogen atoms so liberated are taken up, not by NAD, but by a flavoprotein carrier (possibly FMN or FAD).

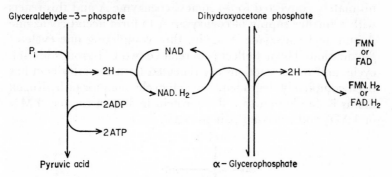

Glyceraldehyde –3– phospate Dihydroxyacetone phosphate

P_i 2H NAD NAD.H_2 2ADP 2ATP 2H FMN or FAD FMN.H_2 or FAD.H_2

Pyruvic acid α– Glycerophosphate

7.2. Fats Fats consist of a higher alcohol, glycerol, combined with up to three fatty acids. They are broken down to these two basic constituents for digestion and are mobilised in the body in this form. Both enter the pathway followed by carbohydrates in their degradation to produce energy.

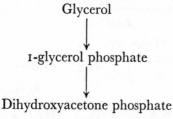

Fat

Glycerol Fatty acids

Glycerol is phosphorylated to form 1-glycerol phosphate, which is then converted to the triose phosphate, dihydroxyacetone phosphate. The net production of ATP is in the region of 26 molecules for each molecule of glycerol.

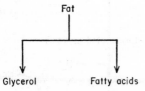

Glycerol

1-glycerol phosphate

Dihydroxyacetone phosphate

The pathway followed by the fatty acids is more complex. They are progressively degraded by a series of reactions known as the β-oxidation or fatty acid cycle. Each fatty acid reacts with coenzyme A to produce acyl coenzyme A, a reaction requiring energy in the form of ATP. The acyl coenzyme A is ultimately converted to β-ketoacyl coenzyme A and this reacts with a further supply of coenzyme A to form acetyl coenzyme A and acyl coenzyme A again, thus completing one cycle of β-oxidation. The net effect is to remove two CH groups in each cycle, and these reactions are repeated until the fatty acid has been completely degraded. In each cycle one pair of hydrogen atoms is handed on to a flavoprotein hydrogen-carrier (FMN or FAD) and a second pair to NAD.

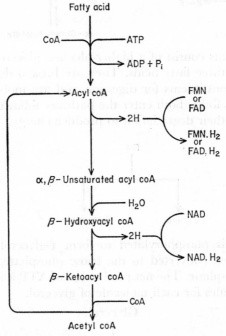

The overall reactions are

$$\text{Fatty acid} \begin{pmatrix} \text{with } n\text{CH}_2 \\ \text{groups} \end{pmatrix} + \text{ATP} + \text{CoA} \rightarrow$$

$$\text{Acyl coA} + \text{ADP} + \text{P}_i$$

$$\text{Acyl coA} + \frac{n}{2}\,\text{CoA} + \frac{n}{2}\,\text{H}_2\text{O} \rightarrow$$

$$\frac{n+2}{2}\,\text{Acetyl coA} + n\,\text{H}_2$$

For example Palmitic acid contains 14 CH_2 groups and thus

Palmitic acid + ATP + CoA→Palmityl coA + ADP + P_i

Palmityl coA + 7 CoA + 7 H_2O→8 Acetyl coA + 14 H_2

and so the degradation of this fatty acid will produce 137 molecules of ATP.

Tripalmitin is a simple triglyceride fat containing one molecule of glycerol combined with three molecules of palmitic acid. The degradation of a molecule of this fat would produce more than 400 molecules of ATP. The overall reaction for palmitin is

$$C_{51}H_{98}O_6 + 72.5\ O_2 → 49\ H_2O + 51\ CO_2$$

The respiratory quotient for a fat, as given by this reaction, is 51/72.5, i.e. between 0.70 and 0.71.

Most invertebrates are thought not to utilise fats to any extent as a store of energy, but to rely on carbohydrates and, in times of stress, on tissue proteins. However, insects certainly do store triglycerides.

7.3. Proteins Proteins are composed of amino acids, of which there are about 20 different ones used in the synthesis of animal proteins. Proteins composed only of amino acids are called simple proteins; others contain additional substances and are known as conjugated proteins. Both are reduced to their constituent amino acids as an initial step in the degradation of the molecule.

The majority of amino acids lose their amino group to form one of the keto acids of the TCA cycle (pyruvic, oxaloacetic, α-ketoglutaric). This may be effected by transferring the amino group to another keto acid to form a different amino acid (transamination). Although many different transaminases are present the most widespread reaction of this type in insects is

Aspartic acid + α-ketoglutaric acid $\rightleftharpoons$ Oxaloacetic acid
+ Glutamic acid

However, this does not reduce the amount of respiratory substrate, but changes its form, and thus is of little or no direct importance in the production of energy.

The degradation of amino acids occurs by a process of deamination, which results in the removal of the amino groups from the molecule. Deamination can occur in one of two ways, depending on the amino acid and hence the enzyme involved. Some undergo an oxidative deamination; others such as glutamic and aspartic acids undergo a hydrolytic deamination, in which a pair of hydrogen atoms is released and passed to the electron transport system via the hydrogen-carrying enzyme NAD. One of the products of this deamination process is ammonia. This is excreted as such, or after conversion to a less soluble substance such as urea or uric acid. The other product is usually a keto acid or one of the other intermediate compounds of the TCA cycle, which one again depending on the amino acid involved. In a few the point of entry into the TCA cycle is via acetyl coenzyme A.

Proteins are not oxidised completely in the tissues, but the respiratory quotient can be found indirectly and lies between 0.80 and 0.82.

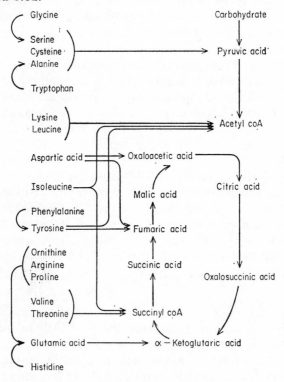

EIGHT
Control Mechanisms

The initiation of the heart beat and nervous control over its regularity, frequency and amplitude have been dealt with in chapter 6. These parameters are of course important in respiration, since the speed with which oxygen reaches the tissues and with which carbon dioxide is removed from them generally depend on the efficiency of the circulatory system. Thus, not only is some efficient pumping organ, the heart, necessary to meet the respiratory demands of the animal, but the closer the immediate requirements can be reflected by alterations in the velocity of blood flow, the more independent from the environment can the animal become, with the proviso that a very high rate of heart beat is clearly undesirable because of the energy required to maintain it.

This chapter is concerned more directly with the mechanics of providing a flow of air or water (depending on the environment) over the respiratory surfaces, that is with the control of ventilatory movements. Detailed information on the neural mechanisms involved and their control is almost exclusively confined to the insects. It can conveniently be categorised into two parts: ventilatory pumping movements of parts of the body —usually the abdominal sterna; and spiracular control and the synchronisation of spiracular movements with ventilatory pumping. These will now be considered in turn.

8.1. Ventilation Many insects, such as the locust and adult dragonfly, ventilate more or less continually, while others do so only intermittently. Thus dragonfly larvae show long periods of ventilation, interrupted occasionally by periods of quiescence. *Byrsotria*, the Cuban burrowing cockroach, normally ventilates between three and fifteen times at a rate of five per minute and then pauses for several minutes. Other cockroaches and also wasps tend to ventilate only during, and for a short period after, activity.

The aquatic dragonfly larva has a closed tracheal system (chapter 4) with fine branches from the main tracheal trunks forming loops in the abdominal gills. The gills are enclosed in a modified region of the hind-gut known as the branchial chamber (figure 8·1). Water is drawn into the branchial

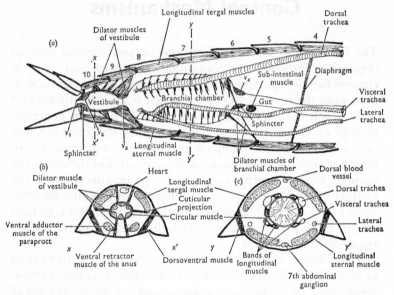

FIGURE 8·1 Longitudinal (*a*) and transverse (*b,c*) sections through the abdomen of the larva of an aeshnid dragonfly to show the branchial chamber and the muscles involved in ventilation. [From Hughes and Mill (1966) *J. exp. Biol.*, **44**]

chamber via the anus (inspiration); oxygen is removed from the water into the tracheal gills and the water is then expelled again through the anus (expiration).

In *Aeshna* and *Anax* the branchial chamber is not very muscular. It possesses a thin investment of circular muscle and six narrow bands of longitudinal muscle. Between its hind end and the anus is a small compartment, the vestibule, with very muscular walls on which a number of dilator muscles insert from the body wall (figure 8·1). There are valves at both ends of the branchial chamber and between the vestibule and anus. Apart from the intrinsic musculature of the branchial chamber and vestibule some of the body wall muscles are also involved in ventilation. In the posterior segments of the abdomen the dorsoventral musculature on each side is subdivided into three separate muscles, the centre one of which acts in an expiratory capacity (expiratory dorsoventral muscle). In the dorsoventrally flattened larva of the genus *Libellula* a second pair of segmental muscles, the oblique tergopleural muscles, are also important in expiration. Two large transverse muscles are associated with inspiration. One, the diaphragm, lies at the anterior end of the fifth abdominal segment and divides the abdomen into two separate compartments. The other lies at the anterior end of the sixth abdominal segment below the intestine (subintestinal muscle) (figure 8·1).

In the resting non-ventilating dragonfly larva the branchial chamber is full of water, and when the animal starts to ventilate, water is ejected through the anus. In other words the initial phase of a normal ventilatory cycle is expiration. The branchial chamber is then re-filled (inspiration) and the cycle is repeated rhythmically at a frequency generally between 25 and 45 cycles/minute in *Aeshna* and *Anax* and somewhat higher in *Libellula* (55–90/minute). In order to illustrate the mechanisms involved, the details of a single cycle will now be considered.

The sterna are raised by the expiratory dorsoventral muscles, thus reducing the abdominal volume and so increasing the pressure in the abdominal cavity. This increase in pressure is transmitted to the branchial chamber and vestibule where it is probably enhanced by contraction of the intrinsic musculature of these chambers. It reaches two to five centimetres of water in the branchial chamber (figure 8·2). At the start of this expiratory phase the anal valve is opened to about one third of its maximum (figure 8·3; plate VI). The pressure reaches a peak and then falls to zero as water continues to be ejected through the anus. At this point the sterna are fully raised and the expiratory phase is complete.

Lowering of the sterna is effected by contraction of the diaphragm and subintestinal muscle as well as by the natural cuticular elasticity of the sclerites, and is thought to be accompanied by contraction of the vestibular dilator muscles. Also, as the sterna start to fall the anal valve opens fully. The effect

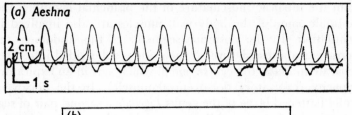

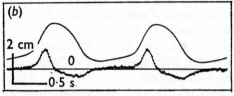

FIGURE 8·2 Dragonfly larva. Records of sternal movement (upper trace—upwards indicates upward movement) and pressure in the branchial chamber with respect to the outside zero pressure (lower trace). [From Hughes and Mill (1966) *J. exp. Biol.*, **44**]

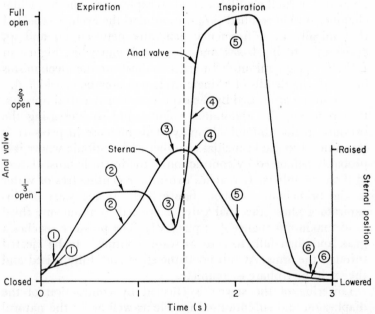

FIGURE 8·3 (*a*) Dragonfly larva. The movement of the sterna and the open anal valve during a single ventilatory cycle. (See plate VI.) [After Mill and Pickard (1972) *J. exp. Biol.*, **56**]

of these actions is to produce a slight negative pressure in the branchial chamber (0.5–1.0 cm H_2O) and so water is drawn in again. The anal valve then closes until the commencement of the next cycle. These changes are summarised in figure 8·10. Increase in frequency of ventilation is largely effected by a reduction in the duration of the resting phase.

During expiration the comparatively high pressure in the branchial chamber combined with the narrow anal aperture causes the oxygen-depleted water to be ejected some distance

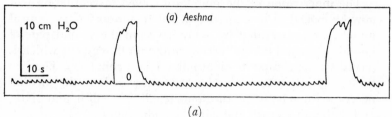

(a)

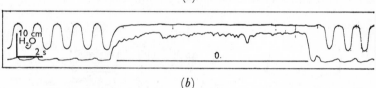

(b)

FIGURE 8·4 Dragonfly larva: gulping and chewing ventilation. (a) Record of pressure changes in the branchial chamber of *Aeshna*; (b) records of sternal movement (upper trace—upwards indicates upward movement) and pressure in the branchial chamber with respect to the outside zero pressure (lower trace). The small oscillations in the pressure records are the normal ventilatory cycles.
[From Hughes and Mill (1966) *J. exp. Biol.*, **44**]

from the animal, so that it is not taken in again during inspiration; the fully opened anal valve of the inspiratory phase ensures that the energy expenditure required to draw fresh water into the respiratory system.[1]

The dragonfly larva uses a modification of this ventilatory process for jet-propulsive swimming. This is a much more rapid activity and the pressures developed in the branchial chamber often exceed 30 cm of water. Other dorsoventral muscles are probably involved and there is also some longitudinal telescoping of the abdominal segments.

Periodically the anal valve remains closed with the sterna in the extended position while the pressure within the branchial chamber is increased by the action of the intrinsic musculature alone, and this may be maintained for several seconds. It is known as 'gulping' ventilation and pressures of over 20 cm of water have been recorded (figure 8·4). Slight fluctuations in this pressure generally occur ('chewing' ventilation) and these are thought to reflect movements of water within the branchial chamber which probably effect a change of water in the 'dead-space' of normal ventilation.

The innervation of the muscles involved in ventilation will now be looked at in some detail. The expiratory dorsoventral muscles are innervated by a branch from the second pair of lateral segmental nerves (except in the ninth segment which is served by the composite eighth abdominal ganglion). The two transverse inspiratory muscles are innervated by the median nerves arising between the fourth and fifth, fifth and sixth, and sixth and seventh abdominal ganglia (figure 8·5).

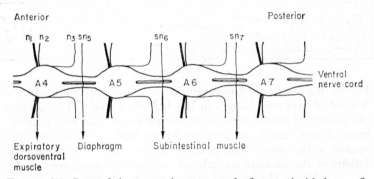

FIGURE 8·5 Part of the ventral nerve cord of an aeshnid dragonfly larva showing the 4th to 7th abdominal ganglia (A4–A7), the segmental nerves (n_1–n_3) and the unpaired nerves (sn_6–sn_8).

Recordings from the second segmental nerves reveal rhythmic bursts of activity coinciding with expiration. Immediately after the end of each expiratory burst an inspiratory burst occurs in the motor neurones innervating the diaphragm and subintestinal muscle (figure 8·6). The expiratory bursts may contain up to three motor units. However, simultaneous recordings from a second segmental nerve and the corresponding expiratory dorsoventral muscle indicate that muscle potentials are only associated with one of these units (figure 8·7); also only a single axon has been seen going to each expiratory dorso-ventral muscle in preparations stained with methylene blue (a vital stain for nerves). The expiratory bursts start first in the last abdominal ganglion, which thus presumably contains a pacemaker, and appears after short delays from successively more anterior ganglia. This is shown in figure 8·8, in which chronic recordings were made from expiratory dorsoventral muscles. The duration of firing and the number of impulses in the expiratory motor neurones decrease from behind forwards and the bursts in different segments cease simultaneously. In each burst the expiratory motor neurone shows a characteristic build up in frequency which reaches its maximum at about the same time in all segments (figure 8·8). This increase in frequency is probably necessary to increase the tone in the muscle, which in turn is needed to overcome the increasing cuticular restoring force which develops as the sterna are lifted; indeed the maximum tension developed by each muscle coincides with the peak frequency (figure 8·9). It is interesting to note that expiratory activity ceases some 100 ms before the sterna are fully lifted. This could be due to the inertia of water flowing out through the anus producing a 'sucking' effect and so continuing the upward sternal movement, or alternatively the expiratory dorsoventral muscles may raise the sterna through a stable position so that they then 'click' into place in the fully raised position. If the latter is the case it explains the need for inspiratory muscles which would force the sterna downwards beyond the position of stability, after which their effect would be enhanced by the natural elasticity of the sclerites. The respiratory system is summarised in figure 8·10.

This rhythm is thought to be initiated and controlled by the central nervous system. Carbon dioxide receptors in the central nervous system probably control the frequency and type of ventilation, while peripheral receptors are not normally important in this respect. Nevertheless, an artificial electrical stimulus applied to one of the first segmental nerves during the period between two expiratory bursts may elicit an expiratory burst in the second segmental nerves and so re-set the phasing of the rhythm (figure 8·11a). Similarly a repetitive series of volleys of electrical stimuli delivered to a first segmental nerve at a frequency slightly higher than that at which ventilation is occurring

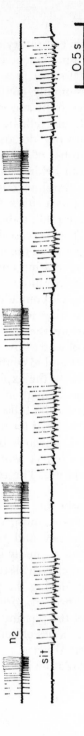

FIGURE 8·6 Dragonfly larva: the alternation of expiratory and inspiratory activity. Expiratory activity (upper trace) is from a second segmental nerve; inspiratory activity (lower trace) is from the subintestinal transverse muscle. [From Mill (1970) *J. exp. Biol.*, **52**]

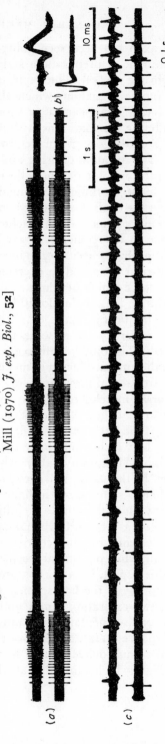

FIGURE 8·7 (*a–c*)

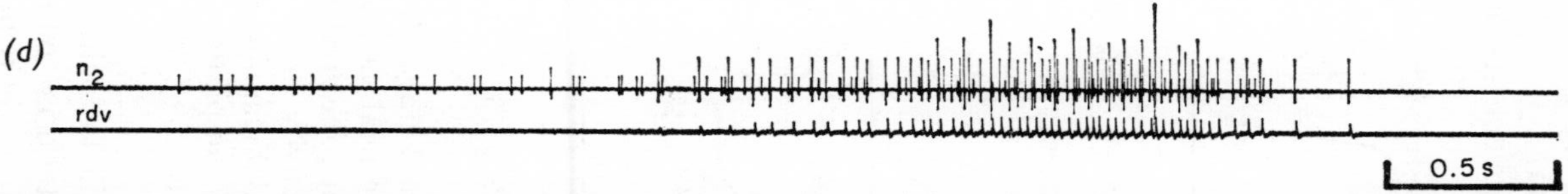

(d)

n₂

rdv

0.5 s

FIGURE 8·7 Dragonfly larva: recordings of activity in a second segmental nerve (upper trace) and the expiratory dorsoventral muscle which it innervates (lower trace). *b* contains superimposed bursts from a single sweep. [*a–c* from Mill and Hughes (1966) *J. exp. Biol.*, **44**. *d* from Mill (1970) *J. exp. Biol.*, **52**]

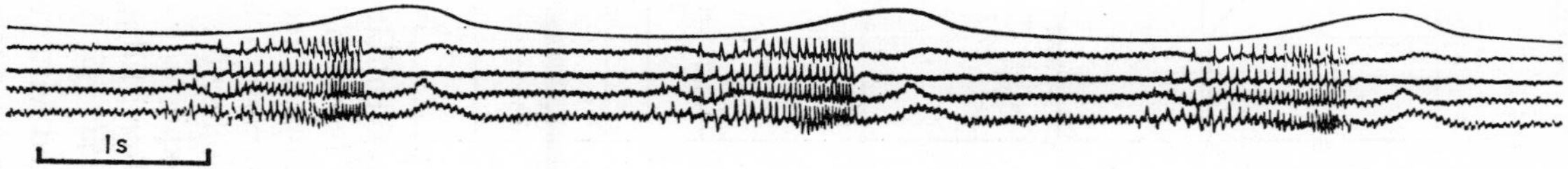

FIGURE 8·8 Dragonfly larva: the upper trace shows sternal movements (upwards indicates upward movement); the other traces are chronic recordings from the expiratory dorsoventral muscles of one side of segments 5 to 8. [From Pickard and Mill. (1972) *J. exp. Biol.*, **56**]

FIGURE 8·9 Dragonfly larva: record of the strain produced by a single expiratory dorsoventral muscle (upper trace) during expiration. Lower trace indicates activity in the appropriate muscle [From Pickard and Mill (1972) *J. exp. Biol.*, **56**]

may cause the rhythm to become phase-locked with the volley frequency; however the original frequency is resumed immediately on cessation of stimulation (figure 8·11*b*).

In the dragonfly larva ventilatory pumping movements of

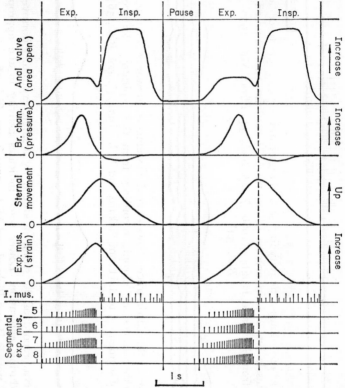

FIGURE 8·10 Summary diagram of ventilation in the aeshnid dragonfly larva showing the correlation between opening of the anal valve, pressure in the branchial chamber, sternal movements, the strain produced by a single expiratory muscle, and the activity in the inspiratory and expiratory muscles. *Br. chamb*, branchial chamber; *Exp.*, expiratory; *Exp. mus.*, expiratory muscle; *I. mus.*, inspiratory muscle; *Insp.*, inspiratory. [After Mill and Pickard (1972) *J. exp. Biol.*, **56**]

the abdomen are clearly linked with ventilation of the branchial chamber. Many insects with an open tracheal system, i.e. one with functional spiracles, use similar ventilatory movements to effect movement of air in and out of the spiracles—again by

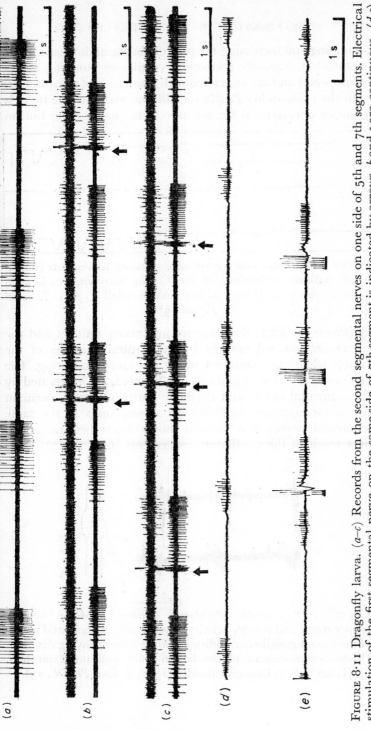

FIGURE 8·11 Dragonfly larva. (a–c) Records from the second segmental nerves on one side of 5th and 7th segments. Electrical stimulation of the first segmental nerve on the same side of 7th segment is indicated by arrows. b and c are continuous. (d,e) Records from the second segmental nerve on one side of 6th segment; the lower trace in e indicates volleys of electrical stimuli delivered to the first segmental nerve on the same side of the 6th segment. d and e are continuous. [a–c from Mill and Hughes (1966) J. exp. Biol., **44**. d, e from Mill (1970) J. exp. Biol., **52**]

decreasing and increasing the pressure in the abdomen; indeed in the adult dragonfly abdominal ventilatory pumping movements persist and are associated with ventilation of the spiracles.

In the cockroaches *Periplaneta*, *Blaberus* and *Byrsotria* there is a simple dorsoventral sternal movement (figure 8·12) but, in

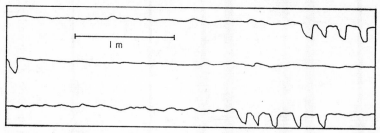

FIGURE 8·12 Continuous records of ventilatory movements (downwards indicate expiration) of the Cuban burrowing cockroach, *Byrsotria fumigata*. [From Myers and Retzlaff (1963) *J. insect Physiol.*, **9**]

Periplaneta, at least, there are no inspiratory muscles and the sterna are lowered entirely by the natural elasticity of the cuticle when the expiratory motor neurones stop firing. The expiratory muscles, as in the dragonfly larva, are innervated by the segmental nerves and rhythmic bursts of activity occur in the motor neurones. In *Periplaneta* sensory units in the same segmental nerves that contain motor axons become active as a result of the ventilatory movements (figure 8·13). Elec-

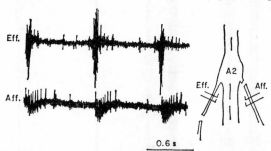

FIGURE 8·13 Cockroach (*Periplaneta americana*). Motor activity (upper trace) and sensory activity (lower trace) in segmental nerves of the second ganglion, as indicated in the accompanying diagram. Downward movement of the botton trace indicates expiration.
[From Farley, Case and Roeder (1967) *J. insect Physiol.*, **13**]

trical stimulation of these nerves has a different effect to that observed in the dragonfly larva, in that the commencement of the next expiratory burst is delayed rather than elicited, and so the ventilatory frequency can be decreased by a repetitive series of volleys delivered at a frequency slightly below that at which ventilation is occurring. As in the dragonfly larva the original frequency is resumed immediately on cessation of stimulation (figure 8·14a). While it could be argued that this difference is due to antidromic (the 'wrong' way) stimulation of the motor neurones in the cockroach, this is negated by the fact that the same effect can be achieved by brief dorsal or ventral movements of the ventral sterna (figure 8·14b), which presumably can only stimulate sensory units. An explanation of this difference must await further studies.

In *Periplaneta* the primary pacemaker is normally situated in the third thoracic ganglion, but it is possible for a centre in the second abdominal ganglion to take over this role under certain conditions. In *Byrsotria* the primary pacemaker is in the first abdominal ganglion.

In the locust *Schistocerca* dorsoventral abdominal ventilation also occurs, but in times of respiratory stress this may be assisted by movements of the head on the prothorax ('neck' ventilation) and of the prothorax on the mesothorax ('prothoracic' ventilation).

Although the basic nature of ventilation in insects is evident, nevertheless there are differences in detail not only between different orders, but between different genera.

8.2. Spiracular movements Among insects there is a tremendous variety in the patterns of opening and closing of the spiracles and their synchronisation with abdominal ventilation. In some, different spiracles are open during inspiration and expiration, thereby producing a unidirectional flow of air through the main tracheal trunks.

In adult dragonflies not only are there generic differences but also individual differences, which depend on a number of factors, maturity for example. However, in flight these differences tend to disappear. When flight commences the first three spiracles immediately open and remain in this position, while spiracles 4–9 are synchronised with ventilation and open only during inspiration. In the locust *Schistocerca* spiracles 3 and 5–9

G

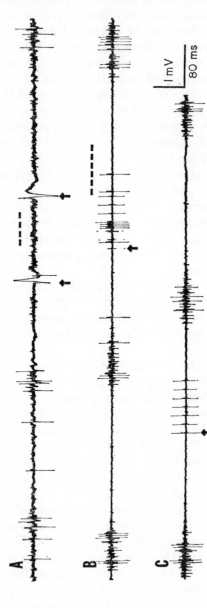

FIGURE 8·14 Cockroach (*Periplaneta americana*). Expiratory bursts in a segmental nerve. (*a*) the sterna were experimentally raised between the two arrows. (*b,c*) Arrows indicate the start of volleys of electrical stimuli delivered to two segmental nerves. The two bursts shown are the first and last of a series. In *a* and *b* the dashed lines indicate the expected occurrence of expiratory bursts.

[From Farley and Case (1968) *J. insect Physiol.*, **14**]

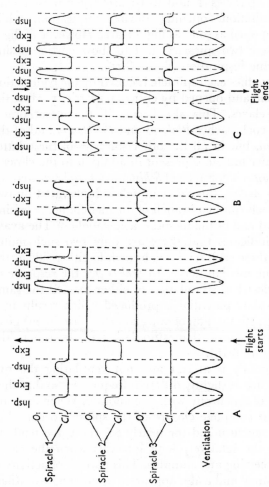

FIGURE 8·15 Locust (*Schistocerca gregaria*). Behaviour of the first three spiracles before, during and after flight. B indicates about 30 minutes after the start of flight. *Cl*, closed; *O*, open; *Ins.*, inspiration; *exp*, expiration. The lowest trace shows the sternal movement (upwards indicates upward movement). [From Miller (1960) *J. exp. Biol.*, **37**]

are closed in the resting animal, while spiracles 1, 2 and 4 open during inspiration and spiracle 10 opens towards the end of expiration (and sometimes in inspiration also). When the animal becomes more active spiracles 5–9 start to open and close. In flight spiracles 1 and 4–10 are synchronised with abdominal ventilation, opening during inspiration, while spiracles 2 and 3 open as soon as flight starts. However, if flight is prolonged these two spiracles start to show partial closing movements during inspiration (figure 8·15).

In adult dragonflies each spiracle is closed by the contraction of a closer muscle and opened by cuticular elasticity when the closer muscle relaxes. This also occurs in the mesothoracic spiracles of the cockroaches *Blaberus* and *Periplaneta* and of the locust *Schistocerca*, but in all the other spiracles of these latter animals an opener muscle is present in addition to the closer.

In the mesothoracic spiracles of *Schistocerca*, *Periplaneta* and the adult dragonfly *Aeshna*, the closer muscle is innervated by two motor axons which run in a branch of the median nerve arising towards the hind end of the mesothoracic ganglion. The available evidence indicates that there are only two such motor axons and that these each send one branch to either side at the bifurcation of the median nerve. In *Schistocerca* the response of the muscle to electrical stimulation of either axon is the same in that an end-plate potential is produced which results in a twitch. However, stimulation of one axon (the 'fast' axon) gives a larger muscle response than stimulation of the other (the 'slow' axon) (figure 8·16).

When a dragonfly is at rest and not ventilating the two motor neurones fire at slightly different frequencies so that they drift in and out of synchrony with one another. When they are not synchronised complete closure of the spiracle occurs, but when they are synchronised (or nearly so) the valve tends to flutter open in the relatively long intervals when the closer muscle is not receiving any stimulus. This pattern of activity is called 'free-running' and it also appears to occur in non-ventilating *Blaberus* and *Periplaneta* (figure 8·17a), and can be experimentally induced in the locust (figure 8·17b).

Periodically superimposed on the free-running pattern there is a pattern associated with expiration. In the adult dragonfly this emanates from a centre in the abdominal part of the central nervous system and the nature of it varies in different genera. It may consist of a high-frequency burst, a cessation of activity, or a high-frequency burst followed by a silent period. In *Blaberus* and *Periplaneta* there are rhythmic bursts synchronised

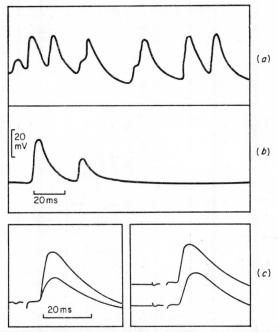

FIGURE 8·16 Locust (*Schistocerca gregaria*). Electrical responses of the mesothoracic spiracle closer muscle to stimulation of the 'fast' and 'slow' motor axons. (*a,b*) Spontaneous activity with fast (large) and slow (small) potentials. (*c*) Two pairs of fast and slow responses, each in the same muscle fibre, elicited by separate electrical stimulation of the axons. [From Hoyle (1969) *J. insect Physiol.*, **3**]

with expiration and for the rest of the cycle the motor units are silent (figure 8·18).

In the locust free-running is not normally observed because ventilatory patterns are superimposed on it most of the time from a centre in the metathoracic ganglion. It does exist though, and can be observed for example in the median nerves to the pro- and mesothoracic spiracles when they are isolated from the metathoracic ventilatory centre by sectioning the nerve cord immediately behind the mesothoracic ganglion. The basic

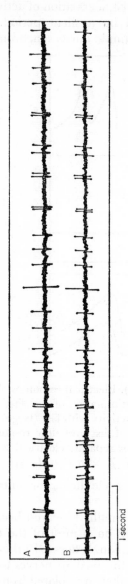

FIGURE 8·17 'Free-running' activity in (a) the closer motor nerves to spiracles 2 (A, right; B, left) of *Periplaneta americana*. There are two closer axons which bifurcate to produce a synchronous output on the two sides. (b) The closer motor nerves to spiracles 1 (SP1) and 2 (SP2) of a locust from which the metathoracic ganglion has been removed. There are two closer axons firing in each nerve. [a from Case (1957) *J. insect Physiol.* **1**. b from Miller (1965) in *The Physiology of the Insect Central Nervous System*. (Ed. Treherne and Beament). Academic Press: London]

1 second

FIGURE 8·18 Rhythmic bursts of expiratory activity in a median nerve of *Blaberus craniifer* during spontaneous hyperventilation. [From Case (1957) *J. insect Physiol.*, **1**] ventilatory pattern to the closer muscles of the pro- and meso-thoracic spiracles in the locust involves an initial high-frequency burst (up to 250 impulses/second/axon) in which there is some synchrony between the units to each spiracle but not between the two spiracles (figure 8·19). This is followed by a reduction in frequency to about 40–60 impulses per second in each axon. During this period there is synchronisation both between the

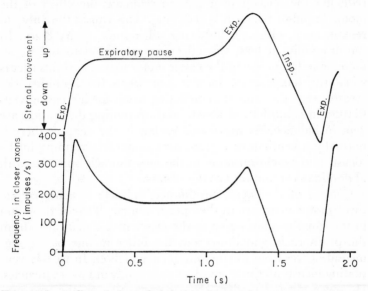

FIGURE 8·19 Correlation of abdominal sternal movements with the frequency of motor impulses to the closer muscle (total of both axons) of the first spiracle of a locust in air. *Exp.*, expiratory; *Ins.*, inspiratory. [After Miller (1965) In *The Physiology of the Insect Central Nervous System*. (Ed. Treherne and Beament). Academic Press: London]

motor units to each spiracle and between the two spiracles. There is then another short burst followed by a period of complete silence. The bursts of activity correspond with upward movement of the abdominal sterna (expiration) and the lower frequency period between the two bursts corresponds with a pause in this upward movement. The silent period is associated with downward movement of the sterna (inspiration). Carbon dioxide acts on centres in the head and thoracic ganglia, and increase in the partial pressure of this gas has the effect of shortening and finally eliminating the expiratory pause (figure 8·21).

Control over the degree of closure in those spiracles without an opener muscle is achieved by the interaction of two antagonistic systems, but is probably only important in animals which are not ventilating, i.e. during free-running activity. Increase in carbon dioxide has a direct effect on the closer muscle, reducing the tension in it and so reducing the effect of the motor impulses which it is receiving. This causes the valve to remain open, although fluttering still occurs (figure 8·20). In the dragonfly at least this effect can be counteracted by an increase in frequency in the closer motor axons, and this occurs when the temperature increases or when the humidity decreases; thus the animal's conflicting needs for increasing ventilation in conditions of anoxia and preventing dessication are kept in balance. As was stated earlier in the chapter there is normally a ventilatory pattern over-riding free-running in the locust, and synchronisation of the mesothoracic spiracles with abdominal ventilation usually occurs.

Control of the locust prothoracic spiracle is achieved via three motor neurones to the opener muscle. Two of these run in the same median nerve as the closer motor neurones (from the prothoracic ganglion), and the other in one of the segmental nerves from the mesothoracic ganglion. In weakly ventilating locusts all three are silent or fire only at low frequencies. However, during moderately strong ventilation the two prothoracic motor neurones normally fire during the latter part of expiration and cease firing at the same time as the closer motor neurones. Nevertheless they still aid opening because the action potentials take longer to reach the periphery than do those in the closer motor axons; also the opener muscle relaxes more slowly than the closer muscle. Opening is enhanced by activity of the mesothoracic neurone, which fires during inspiration. In the presence of carbon dioxide the prothoracic motor neurones continue firing over a longer part of expiration, which is itself shortened under such conditions, and extend into the beginning of inspiration, thus having a more marked effect (figure 8·21). In still higher concentrations all three neurones may fire throughout the cycle.

As was mentioned at the start of this section the pattern of

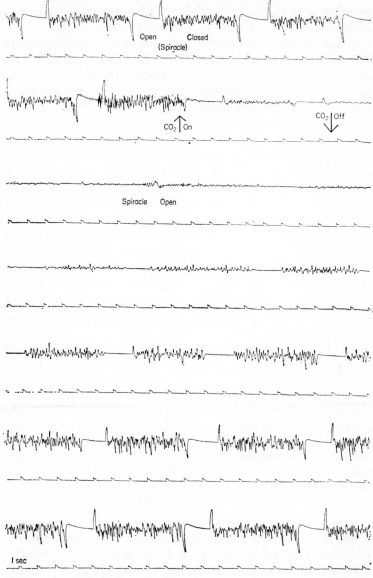

FIGURE 8·20 The effect of CO_2 applied locally to a mesothoracic spiracle of *Schistocerca gregaria*. This is a mechanical record of movements of the lip of the posterior valve. [From Hoyle (1960) *J. insect Physiol.*, **4**]

opening and closing of individual spiracles may change in flight (dragonflies and *Schistocerca*), and one particularly interesting mechanism that occurs in the locust during such activity is worth discussing. The first spiracle in this animal leads into an atrium from which arise two orifices. The dorsal one communicates with tracheae leading to the head and to the longitudinal ventral trunk which supplies the central nervous

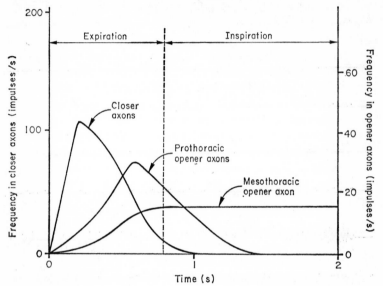

FIGURE 8·21 One ventilatory cycle of a locust in 3% CO_2 to show the frequency of motor impulses to the closer muscle (both axons) and the opener muscle (two prothoracic and one mesothoracic axons shown separately). Note the absence of an expiratory pause with the increased concentration of CO_2. [After Miller (1965) in *The Physiology of the Insect Central Nervous System*. (Ed. Treherne and Beament). Academic Press: London]

system; the ventral one communicates with the pterothoracic tracheal system which is otherwise virtually isolated from the rest of the tracheal system. When the opener and closer muscles contract simultaneously, as during moderately strong ventilation, the ventral orifice is constricted and the isolation of the pterothoracic system maintained. However, the prothoracic opener motor neurones are virtually silent during flight and so the opener muscle is relaxed, allowing air to flow across the atrium from the pterothoracic ventral orifice to the dorsal orifice when the spiracle is closed. If the carbon dioxide content is high centres in the central nervous system stimulate the pro-thoracic opener neurones, further flow across the atrium is prevented and ventilation is adjusted appropriately to deal with the excess carbon dioxide. Thus the central nervous system can periodically sample the pterothoracic gases without being flooded with carbon dioxide.

Opening of the anterior spiracles in dragonflies and some other insects may be effected by mechanical and visual stimuli which normally give rise to flight. It may also occur accom-panying vigorous leg movements. Stimulation of hairs around the thoracic spiracles in *Schistocerca*, *Blaberus*, *Periplaneta* and the adult dragonfly produces a sensory response which enters the central nervous system via the segmental nerves and reflexively elicits closure of both spiracles of the stimulated segment. In the prothoracic segment of *Schistocerca* at least this latter reflex motor response is confined to the slow axon. No sensory activity has been recorded as a result of increase in carbon dioxide tension, and the effect of carbon dioxide is either peripheral (on the tension of the closer muscle) or directly on centres in the central nervous system.

It has been shown that isolated segments are capable of ven-tilatory movements in some insects and attempts have been made to discover the nature of the mechanism underlying the motor output. Rhythmic bursts of activity within the frequency range of normal ventilation have been recorded from the isolated nerve cord of certain insects, sometimes appearing only in the presence of carbon dioxide, but this does not necessarily imply that they are a part of the ventilatory mechanism in the living animal. However, in some cases, recordings from the lateral nerves of isolated nerve cords, or even of individual ganglia, show bursts which are very similar indeed to normal expiratory bursts. This has been demonstrated for example in the locust and in the cockroaches *Byrsotria* and *Blaberus*. In isolated individual ganglia of *Byrsotria* they occur in lateral nerves of the first and sixth abdominal ganglia, but only in the former does carbon dioxide increase the burst frequency. Simi-larly in *Blaberus* the first abdominal ganglion is of prime importance, but in this animal either the third thoracic or second abdominal ganglion appears to be necessary in addition for bursts to appear.

In the locust rhythmic bursts of activity travel posteriorly in the nerve cord from the metathoracic ganglion in the intact animal and probably initiate the expiratory motor bursts in the lateral nerves. In the adults of some species of dragonfly rhythmic bursts ascend the nerve cord to the third thoracic/first abdominal ganglion complex from a centre further back in the abdomen.

The endogenous nature of the ventilatory rhythm seems to be generally accepted insofar as receptors extrinsic to the central nervous system are not involved in its initiation and control; but internal receptors, sensitive to carbon dioxide tension for example, are of importance. The inherent rhythmicity often present in more than one ganglion is normally harnessed by a pacemaker centre, which is itself driven by one or more command interneurones (that is interneurones responsible for driving systems).

OSCILLATING SYSTEMS

There are various ways in which a nervous system could work so as to produce activity first in one set of motor neurones and then in an antagonistic set, i.e. an oscillating system. Some of the possibilities will now be considered, although not all of them have been put forward in connection with work on ventilation. The reader is referred to the bibliography, which includes references to oscillatory systems concerned with cyclical activity other than ventilation.

In figure 8·22a a command interneurone is spontaneously active and fires at a steady frequency. It synapses with two other cells (2 and 3). Initially one of these (2) may have a slightly lower threshold than the other and so starts to fire first. While it is firing it inhibits cell 3 but, as soon as it stops, cell 3 becomes active and in turn inhibits cell 2. In other words there is reciprocal inhibition between cells 2 and 3, which could be motor cells. However, to produce a pause between activity in two antagonistic motor cells it is necessary to introduce further complexity. In figure 8·22a 4 and 5 are the two motor cells and they only fire when they receive a high frequency input, i.e.

during the peak activity of cells 2 and 3 respectively. Alternatively, this same output could be achieved from cells 2 and 3 if they inhibited themselves when they started to fire such that this self-inhibition lasted longer than the reciprocal inhibition (figure 8·22*b*).

Another possibility would be for the regularly firing command interneurone to synapse with only one motor cell (cell 2

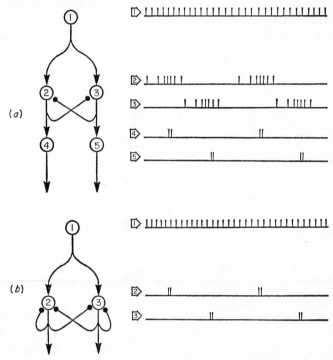

FIGURE 8·22 Hypothetical scheme for an oscillating system. I.
[After Wilson (1964) *J. exp. Biol.*, **41**]

in figure 8·23). In this example cell 2 is self-excitatory and synapses with cell 3, which has a high threshold, only becoming active when the input frequency from cell 2 is high. When this occurs it inhibits cell 2 and excites the other motor cell (4). This would give an oscillating output reminiscent of that occurring in ventilation in the dragonfly larva, where cell 2 would be an expiratory and cell 4 an inspiratory motor neurone.

In the system shown in figure 8·22 alterations in frequency

of the oscillating output could possibly be achieved by varying the frequency in the command interneurone. However, the nature of cells 2 and 3 would tend to impose stability on the frequency of the system. For example, an increase in frequency in the command interneurone might increase the number of impulses in the output of cell 2, but any such change would increase the inhibition of cell 3, which in turn would tend to counteract the increased excitatory effect of the command interneurone on this cell, and so on. This would be a very useful system where stabilisation of the frequency of oscillation is

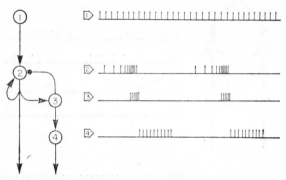

FIGURE 8·23 Hypothetical scheme for an oscillating system. II. [Based on Hughes and Mill (1966) *J. exp. Biol.*, **44**]

important. However, such changes could be achieved with the arrangement shown in figure 8·22 if the output from cells 2 and 3 is independent of the absolute frequency of the command interneurone. Under such conditions the inhibition would always be the same, while the expiratory input to the cells could be varied. In figure 8·23 changes in frequency of oscillation are readily brought about by changes in frequency of the command interneurone. Thus, as cell 2 receives a greater frequency of input its output frequency will increase and so cause cell 3, and hence cell 4, to fire sooner.

The activity in the command interneurone itself could wax and wane due to intrinsic oscillations in its own membrane potential. Figure 8·24a shows motor neurones (2 and 3) innervated sequentially by the command interneurone. They fire when their input rises above a certain threshold frequency level. The main difficulty here is the very short delay between (*a*)

and (b). This could reach an acceptable value if the antagonistic motor neurones are in different ganglia, otherwise it is
necessary to envisage additional steps in the system, such as the
interpolation of a relaying interneurone. Changes in frequency
of the oscillating output can readily be achieved by changes in
the frequency of the activity wave in the command interneurone (figure 8·24b). Furthermore, changes in the amplitude

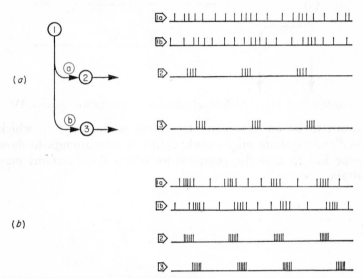

FIGURE 8·24 Hypothetical scheme for an oscillating system. III.
[Based on Davis (1969) *J. exp. Biol.*, **50**]

of this waxing and waning in the command interneurone could
affect the firing pattern of the motor cells (figure 8·24b).

Intrinsic oscillations in the membrane potential of the command interneurone could cause regular bursts of activity
(figure 8·25). In this example the two motor neurones (2 and
3) are spontaneously active and continue to fire regularly in
the absence of any input. However, cell 2 is inhibited when it
receives an input from the command interneurone and also
activity in cell 2 inhibits cell 3. This resembles the situation in
ventilation and spiracle control in the locust, where free-running in the spiracular closer motor neurones occurs when they
are isolated from the metathoracic pacemaker. Cell 2 could be

a closer or an expiratory motor neurone, cell 3 an opener or an inspiratory motor neurone.

It must be pointed out to the reader that this is in no way

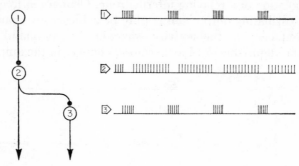

FIGURE 8·25 Hypothetical scheme for an oscillating system. IV

meant to be an exhaustive treatment of the ways in which oscillatory systems might work; rather it is an attempt to show some insight into the complexities which these systems may attain.

Classification

The classification used in this book follows that of Meglitsch, which is a modification of the one suggested by Hyman. All the animals mentioned in the text belong to the Eumetazoa, Bilateria.

DIVISION PROTOSTOMIA

SUBDIVISION ACOELOMATES
  Phylum Platyhelminthes
  Phylum Nemertinea (Rhyncocoela)
SUBDIVISION PSEUDOCOELOMATES
  Phylum Nematoda
SUBDIVISION COELOMATES
  Phylum Phoronida
  Phylum Brachiopoda
  Phylum Sipunculoidea
  Phylum Mollusca
      Class Monoplacophora
      Class Polyplacophora
      Class Aplacophora
      Class Scaphopoda
      Class Gastropoda
      Class Bivalvia (Pelecypoda)
      Class Cephalopoda
  Phylum Echiuroidea
  Phylum Annelida
      Class Polychaeta
      Class Oligochaeta
      Class Hirudinea
  Phylum Onychophora
  Phylum Arthropoda
      Class Merostomata
      Class Diplopoda
      Class Symphyla
      Class Chilopoda
      Class Arachnida
      Class Insecta
      Class Crustacea

DIVISION DEUTEROSTOMIA
    Phylum Echinodermata
        Class Crinoidea
        Class Holothuroidea
        Class Echinoidea
        Class Asteroidea
        Class Ophiuroidea
    Phylum Pogonophora
    Phylum Hemichordata
        Class Enteropneusta
        Class Pterobrancia
    Phylum Chordata
        Subphylum Urochordata

Bibliography

In the main body of the book the figures have been carefully chosen to be representative of the specific literature, and details of their origins are given in the appropriate legends. There follows a list which includes more general texts as well as reviews of certain areas of study, and these have nearly all been published within the last decade.

General

Barnes, R. D. (1968) *Invertebrate Zoology*, 2nd Ed. Saunders: Philadelphia

Chapman, R. F. (1969) *The Insects: Structure and Function*. English Universities Press: London

Grassé, P.-P. *Traité de Zoologie*. Vols. IV–XI. Masson: Paris

Hughes, G. M. (1963) *Vertebrate Respiration*. Heinemann: London

Hyman, L. H. *The Invertebrates*. Vols. II and IV–VI. McGraw Hill: New York

Jones, J. D. (1963) In *Problems in Biology*. Vol. I. (Ed. G. A. Kerkut). Pergamon Press: Oxford

Meglitsch, P. A. (1967) *Invertebrate Zoology*. Oxford Univ. Press: New York

Newell, R. C. (1970) *Biology of Intertidal Animals*. Logos: London

Nicol, J. A. C. (1967) *The Biology of Marine Animals*. 2nd Ed. Pitman and Sons: London

Prosser, C. L. and Brown, F. A. (1961) *Comparative Animal Physiology*. 2nd Ed. Saunders: Philadelphia

Rockstein, M. (Ed.) (1964) *The Physiology of Insecta*. Vol. III. Academic Press: New York. 2nd Ed., Vol. IV (In the press)

Waterman, T. H. (Ed.) (1960) *The Physiology of Crustacea*. Vol. I. Academic Press: New York

Wigglesworth, V. B. (1965) *The Principles of Insect Physiology*. 6th Ed. Methuen: London

Wilbur, K. M. and Yonge, C. M. (Eds.) (1966) *Physiology of Mollusca*. Vol. II. Academic Press: New York

Index